SCIENTIFIC MATERIALISM

EPISTEME

A SERIES IN THE FOUNDATIONAL,

METHODOLOGICAL, PHILOSOPHICAL, PSYCHOLOGICAL,

SOCIOLOGICAL AND POLITICAL ASPECTS

OF THE SCIENCES, PURE AND APPLIED

Editor: MARIO BUNGE

Foundations and Philosophy of Science Unit, McGill University

Advisory Editorial Board:

VOLUME 9

MARIO BUNGE

Foundations and Philosophy of Science Unit, McGill University

SCIENTIFIC
MATERIALISM

D. REIDEL PUBLISHING COMPANY

DORDRECHT : HOLLAND / BOSTON : U.S.A.

LONDON : ENGLAND

Library of Congress Cataloging in Publication Data

Bunge, Mario Augusto.
    Scientific materialism.

    (Episteme ; v. 9)
    Bibliography : p.
    Includes indexes.
    1. Science – Philosophy.   2.   Materialism.   I.   Title.   II.   Series:
Episteme (D. Reidel) ; v. 9.
Q175.B827   501   81–10653
ISBN 90–277–1304–9   AACR2
ISBN 90–277–1305–7 pbk.   (Pallas edition)

Published by D. Reidel Publishing Company,
P.O. Box 17, 3300 AA Dordrecht, Holland.

Sold and distributed in the U.S.A. and Canada
by Kluwer Boston Inc.,
190 Old Derby Street, Hingham, MA 02043, U.S.A.

In all other countries, sold and distributed
by Kluwer Academic Publishers Group,
P.O. Box 322, 3300 AH Dordrecht, Holland.

D. Reidel Publishing Company is a member of the Kluwer Group.

Printed in The Netherlands

To my children

ERIC RUSSELL and SILVIA ALICE

May their generation enjoy and improve
the only world we've got
and which my generation may still destroy.

TABLE OF CONTENTS

PREFACE

The word 'materialism' is ambiguous: it designates a moral doctrine as well as a philosophy and, indeed, an entire world view. Moral materialism is identical with hedonism, or the doctrine that humans should pursue only their own pleasure. Philosophical materialism is the view that the real world is composed exclusively of material things. The two doctrines are logically independent: hedonism is consistent with immaterialism, and materialism is compatible with high minded morals. We shall be concerned exclusively with philosophical materialism. And we shall not confuse it with realism, or the epistemological doctrine that knowledge, or at any rate scientific knowledge, attempts to represent reality.

Philosophical materialism is not a recent fad and it is not a solid block: it is as old as philosophy and it has gone through six quite different stages. The first was ancient materialism, centered around Greek and Indian atomism. The second was the revival of the first during the 17th century. The third was 18th century materialism, partly derived from one side of Descartes' ambiguous legacy. The fourth was the mid-19th century "scientific" materialism, which flourished mainly in Germany and England, and was tied to the upsurge of chemistry and biology. The fifth was dialectical and historical materialism, which accompanied the consolidation of the socialist ideology. And the sixth or current stage, evolved mainly by Australian and American philosophers, is academic and nonpartisan but otherwise very heterogeneous.

Ancient materialism was thoroughly mechanistic. Its great names were Democritus and Epicurus as well as Lucretius. Seventeenth-century materialism was mainly the work of Gassendi and

Hobbes. Eighteenth-century materialism, represented by Helvetius, d'Holbach, Diderot, La Mettrie, and Cabanis, exhibited a greater variety. Thus while La Mettrie regarded organisms as machines, Diderot held that organisms, though material, possess emergent properties. The 19th century "scientific" materialists, while philosophically naive, had the merit of linking materialism to science, though not to mathematics. Not only the scientists Vogt, Moleschott and Czolbe were among them but also Tyndall and Huxley and, secretly, Darwin as well. Dialectical materialism, formulated mainly by Engels and Lenin, was dynamicist and emergentist, and claimed to be scientific while at the same time being committed to an ideology. Finally the newer or academic materialists come in a variety of shades, from physicalists like Neurath, Quine and Smart to emergent materialists like Samuel Alexander and Roy Wood Sellars. Their relationship to contemporary science is remote.

Most philosophers from Plato onwards have dismissed philosophical materialism as crass and incapable of accounting for life, mind, and their creations. Accordingly materialism is seldom discussed in the philosophical literature and in the classroom except when allied to dialectics. As a result materialism is still in its infancy even though it is several thousand years old.

Philosophical materialism has been attacked on several counts. Firstly for conflicting with the magical and religious world views. (For this reason it is often conflated with positivism.) Secondly because the dialectical version of materialism is part of the Marxist ideology and therefore anathema (when not untouchable dogma). Thirdly for having allegedly failed to solve the major philosophical problems, or even for having dodged some of them altogether. We shall not be concerned with the first two criticisms for being ideological not philosophical.

We shall tackle instead the philosophical objection that materialism is insignificant for not facing, let alone solving, some of the key problems of philosophy. Here are some of the outstanding

problems that materialism is supposedly unwilling or even incapable
of tackling:

> (i) *How can materialists hold the fort in the face of
> the apparent dematerialization of the world accomplished
> by contemporary physics, with its fields and probability
> waves?*
>
> (ii) *How can materalism, which is supposedly reduc-
> tionistic, explain the emergence of new properties, in
> particular those peculiar to organisms and societies?*
>
> (iii) *How can materialism explain mind, which is im-
> material?*
>
> (iv) *How is materialism to account for purpose and
> freedom, which so obviously transcend natural law?*
>
> (v) *How do materialists make room for cultural
> objects, such as works of art and scientific theories,
> which seem to dwell in a realm of their own and obey
> supraphysical laws or perhaps none at all?*
>
> (vi) *How do materialists propose to explain the causal
> efficacy of ideas, in particular the technological and
> political ones?*
>
> (vii) *Since concepts and propositions have no physical
> properties, how could they possibly dwell in a purely
> material world?*
>
> (viii) *Since the truth of mathematical and scientific
> propositions does not depend on the knowing subject or
> his circumstances, how can it possibly be explained in
> terms of matter?*
>
> (ix) *How can materialism account for values, which
> are not physical entities or properties, and yet guide
> some of our actions?*
>
> (x) *How can materialism explain morality without
> endorsing hedonism, given that the rules of moral be-*

havior, particularly those concerning duties, are alien to natural law?

It must be owned that most materialists have not proposed satisfactory answers to the above crucial questions. Either they have not faced some of them or, when they have, their answers have tended to be simplistic, such as the theses that spacetime points are just as real as chunks of matter, that there is no mind, and that mathematical objects are just marks on paper. In particular, there seem to be no full fledged materialist philosophies of mind and of mathematics, or of values and morals.

To be sure not all materialists are vulgar or crass, and a number of materialist philosophers have offered valuable insights into the above questions. Still, most materialist philosophers speak only ordinary language – and so are bound to formulate their views in an inexact fashion – and they seldom care to argue for them in a cogent way. Besides, materialists have been so busy defending themselves from ignorant or vicious attacks, and counterattacking, that they have neglected the task of building comprehensive philosophical systems and moreover systems compatible with contemporary logic, mathematics, science, and technology. As a result materialism is less a field of active research teeming with novelty than a body of belief, much of it obsolete or irrelevant. (When did you last hear of a recent breakthrough in materialist philosophy?)

While all of this is true, the interesting question is whether materialism is hopelessly dated and impotent, or can be revitalized and updated and, if so, how. This is the problem the present book addresses. This book can be regarded as an invitation to look at materialism as a field of research rather than a body of fixed beliefs. More precisely, it is a challenge to examine, clarify, expand and systematize materialism in the light of contemporary logic, mathematics and science rather than in that of the history of ideas, let alone that of political ideology. Materialism must take up this gauntlet under penalty of remaining underdeveloped and thus uninteresting and inefficient.

The motivations for this challenge as the following. First, materialism has not advanced far beyond the 19th century, partly for having ignored modern logic and refused to learn from rival philosophies. And yet it can be argued that materialism is not just one more ontology: that it is the ontology of science and technology. In particular, materialism is the ontological driving force behind certain scientific breakthroughs such as atomic and nuclear physics, evolutionary biology, the chemical theory of heredity, the scientific study of the origin of life, the physiology of ideation, and the most recent advances in paleoanthropology and historiography.

A second motivation is the author's conviction that philosophical investigation should be conducted systematically, exactly, and scientifically rather than in the manner of literature, let alone pamphleteering. Part of this belief is the thesis that, whereas a philosophical doctrine can be destroyed by analysis or argument, it is best established by showing that it harmonizes with science and that it helps advance scientific research rather than block it. If this be scientism, let it be so.

A third motivation is the thesis that the usual relation of philosophy and ideology, where the former is ancillary to the latter, should be inverted. An ideology cannot be both true and effective unless it agrees with both philosophy and science, which are advanced only by the free search for truth. (What is sometimes called 'the ideology of science' is not an ideology proper but a collection of ontological, epistemological and moral principles concerning reality and the ways of knowing about it.) In particular, materialism should not be rejected or embraced just because it agrees or disagrees with a given ideology. Thus, whether or not we think with our brains is a problem of great ideological import, but not one to be solved by ideology.

Intellectual challenges are self-challenges in the first place. So, this book is more than a challenge to fellow philosophers and scientists: it is also an attempt to sketch solutions to some of the outstanding problems listed a while ago. These solutions are offered tentatively as embryos that may deserve to grow through further research. Some of them have already been developed into full fledged theories to be found in the author's *Treatise on Basic Philosophy* (Bunge, 1974a, 1974b, 1977a, 1979, and forthcoming) and *The Mind-Body Problem* (Bunge, 1980). However, no philosophical system, even if exact and up to date, can be expected to be impervious to criticism and ulterior development or even replacement. Philosophizing may be perennial, philosophies not.

Finally, a warning to the reader who expects to find in this book a review of the various materialist schools, or at least an exposition of a well known materialist philosophy, such as physicalism, or dialectical materialism. He will find neither. What he will find is a sketch of a new ontology built in response to the problems listed above, which were left unsolved by the traditional materialisms. This new ontology will be called *scientific materialism* because it draws its inspiration from science and is tested as well as modified by the advancement of science.

Montréal, Qué., Canada   MARIO BUNGE
February 1981

PART ONE

BEING

CHAPTER 1

MATTER TODAY

Nonmaterialists have a low opinion of matter and one that is not countenanced by the sophisticated theories of matter elaborated by contemporary science. Let us make a quick review of some views on matter still popular in philosophical quarters.

1.1. MATTER INERT?

The most ancient of such views and one that still has partisans is Plato's. According to it matter is the passive receptacle of forms (properties), which in turn are ideas: only the soul (or mind) is self-moving. This was not Aristotle's view, according to whom forms, far from being prior to matter and entering it from the outside, were generated by matter itself. In particular the soul, rather than being self-existing and detachable from the body, was to Aristotle the form of the latter.

From Antiquity onwards, all materialists have held that change is essential for matter. Even though the ancient materialists regarded the atoms themselves as unalterable, they supposed them to be forever in motion. And even though the 18th and 19th century materialists usually regarded force as extrinsic to matter and the cause of the latter's changes in state of motion, they held that no bit of matter can be forever free from the action of forces. Materialism, in short, has always been dynamicist (though only occasionally dialectical). The thesis of the passivity of matter is typically non-materialist.

The dynamicist conception of matter has also been that of the physicists and chemists since Galileo, Descartes, and Boyle. In particular, Newton's principle of inertia states, in opposition to Aristotle's physics, that a body, once in motion, continues to move by itself unless it is stopped by an external force. And both the corpuscular and the wave theories of light assumed that light propagates by itself without being pushed: it is self moving. (Kant, who could not read Newton's equations for lack of mathematical knowledge, misunderstood Newtonian physics as asserting that whatever moves does so under the action of some force, be it attractive or repulsive. And Voltaire, who did so much for the popularization of Newtonian physics in his Cartesian country, was struck by the pervasiveness of gravitation but could not understand it adequately because he, too, was unable to read Newton's equations of motion. So neither Voltaire nor Kant realized that the inertia of bodies and light refutes the belief that matter is inert, i.e. incapable of moving by itself.)

Classical physics, in sum, regarded matter — whether of the genus body or the genus field — as essentially active. So much so that the nucleus of every physical theory since Newton is a set of equations of motion or field equations, as the case may be, which describe, explain and predict the motion of particles, the flow of fluids, the propagation of fields, or some other kind of change.

Needless to say, this dynamicist conception of matter was adopted by chemistry. Indeed chemistry studies not only the composition and structure of chemical compounds but also the processes of formation and transformation (in particular dissociation) of of such compounds. So much so that chemical reactions constitute the very core of chemistry. Moreover, as is well known, whereas classical physics ignored qualitative transformations, chemistry specializes in them. The same can be said of biology since Darwin and of social science since Marx, namely that the former is particularly interested in the transformations of living matter and the latter in the transformations of social matter.

Contemporary science has, if anything, stressed the dynamism

of matter as well as its unlimited capacity to generate new forms
— an Aristotelian first. Think of the humble electron which, even
in isolation, is attributed not only a translational motion but also
a spontaneous trembling motion as well as a spin. Or think of the
modest photon, or any other field quantum, travelling relentlessly
until scattered or absorbed by a particle. So, even the elementary
particles and fields are perpetually changing. A fortiori, all material systems are changeable. Think of atoms, molecules, crystals,
fluids, cells, multicellular organisms, social systems, entire societies,
and artifacts: think of the marvellous variety of their properties,
in particular their properties of undergoing or causing change.

From physics to history science seems to study matter of
various kinds and only matter, inanimate or alive, in particular
thinking and social matter. This is surely a far cry from the view
of matter offered by non-materialist philosophers, in particular the
immaterialist (or idealist) ones. The kind of materialism suggested by
contemporary science is dynamicist rather than staticist. It is also
pluralistic in the sense that it acknowledges that a material thing
can have many more properties than just those mechanics assigns
it. More on this below.

True, any sufficiently advanced scientific theory contains some
conservation law or other — e.g. theorems of conservation of the
total mass, or total momentum, or total energy, or what have you.
Such conservation laws have occasionally been interpreted as refuting dynamicism. But this is a blunder, for conservation formulas
state the permanence of some property of a material thing of a
certain kind amidst change. These properties are constants of the
motion or, in general, constants of the transformation of things.
(Trivial example: the difference in age between parent and child
remains constant as long as both stay alive).

In sum, science denies the thesis that matter is inert and sup-
ports instead the philosophical generalization that all matter is
continually in some process of change or other.

1.2. MATTER DEMATERIALIZED?

A second rather widespread opinion is that modern physics has dematerialized matter. (See e.g. McMullin, ed., 1964, in particular the paper by N. R. Hanson.) There are several versions of this view. One is that physics has shown matter to be a set of differential equations, hence an immaterial entity. This thesis rests on a faulty semantics, according to which a scientific theory coincides with its mathematical formalism. Every physicist knows that this is false: that a set of mathematical formulas must be assigned a set of "correspondence rules", or semantic assumptions, in order to acquire a physical content, i.e. in order to describe a physical entity. Thus the formula "$F = q_1 q_2 / \epsilon r^2$" is not Coulomb's law of electrostatics unless one adds the semantic assumptions that 'F' represents the force of interaction between two point particles with electric charges q_1 and q_2, separated by the distance r, and immersed in a medium of dielectric capacity ϵ. In sum, a physical theory is a mathematical formalism together with a physical interpretation. And the theory, far from being identical with its referent (a physical entity), represents or describes it (whether accurately or poorly).

A second version of the dematerialization thesis is that, after all, every physical entity is a field or is reducible to fields; and, since fields are not material, physical entities are not material either. This view might have been defended over a century ago, when the field concept was young and insecure, and seemed to many to be just a convenient way of summarizing information concerning actions among bodies. But since at that time physics did not regard bodies as being ultimately reducible to fields, the view would have been discarded right away. Ever since Maxwell formulated the classical electromagnetic theory, Hertz produced electromagnetic waves, and Einstein divested the theory of the mythical ether, the field concept has come a long way: it is not regarded as a convenient fiction but as representing a real though subtle entity. Shortly before the emergence of the quantum theory matter could have been defined as the union of two genera: bodies (in particular particles) and fields. Since then we have learned to regard particles as quanta of fields of a kind unknown to classical physics. (For example, electrons are quanta of the electron field.) And we analyze bodies into particles and the fields that hold these together. So, fields have become the basic mode of matter.

A third version of the dematerialization thesis is based on the Copenhagen interpretation of the quantum theory. According to that interpretation, this theory is not about independently existing physical entities but about experimental set-ups that include experimenters. Every quantum event would then be ultimately the result of arbitrary decisions made by a human subject. The theory, which is highly accurate, would then concern matter-mind compounds. Moreover, the line between the material and the mental component can be drawn arbitrarily by the experimenter himself, so there is no objectively or absolutely existent matter. So far the Copenhagen interpretation, which has been subjected to severe criticisms (e.g., Bunge, 1955; Popper, 1967; Bunge, 1973a, b).

One flaw of this interpretation is that no formula of the theory contains variables describing any properties of human subjects, in particular psychological properties. (Note especially that the total energy operator does not contain any contributions from the subject.) Another defect is that many experiments can be automated to the point that their outcomes can be printed and read out by the experimenter after they are completed, which is a way of guaranteeing the subject's non-intervention in the process. So, the quantum theory does not support at all the thesis that matter has been spiritualized.

Finally, a fourth version of the dematerialization thesis is the claim that modern physics has shown the world to be composed of events not things. This belief betrays superficiality, for it is not preceded by an analysis of the concept of an event. In fact, by definition an event is a change of state of some thing or material entity: there are no events in themselves but only events in some thing or other, be it body or field or any other material object. So much so that the simplest analysis of the concept of an event is this: "x is an event in thing y relative to reference frame $z =_{df}$ i and f are possible states of thing y relative to reference frame z, and x equals the ordered pair $\langle i, f \rangle$". Physics does not view the world as composed of immaterial events or of unchanging material objects: the world of physics is a system of changing things, namely the most comprehensive system of this kind.

In conclusion, the rumour that contemporary physics has dematerialized matter turns out to be false. Rather, as we shall see in a moment, physiological psychology has materialized mind.

1.3. LIFE IMMATERIAL?

Vitalism, a descendant of animism, holds that life is the immaterial entity animating organisms and that the latter are designed so that they can achieve their purpose, which is the preservation of their kind. According to materialism, on the other hand, life is a property of certain material objects. To be sure mechanistic materialism denies that there is any qualitative difference between organisms and nonliving things: that the difference is only one of complexity. This kind of materialism is an easy prey to vitalism, for a modern factory is no less complex than a cell, and it is plain that biology studies a number of properties and processes unknown to physics and chemistry. So, mechanistic materialism is not an answer to vitalism.

A materialist conception of life has got to acknowledge emergence, i.e. the fact that systems possess properties absent from their components. In particular biosystems are capable of maintaining a fairly constant internal milieu, the activities of their various parts are coordinated, they can self-repair to some extent, they can reproduce, cooperate, and compete, and they are subject to evolution. Emergentist materialism has no trouble acknowledging the peculiarities of biosystems. Moreover, unlike holism, emergentist materialism encourages the search for an explanation of emergence in terms of lower level properties and processes.

How do vitalism and emergentist materialism fare in modern biology? The answer depends on the kind of textual evidence one selects, for whereas some texts favor vitalism others defend mechanism (or physicalism) while still others tacitly endorse emergentist materialism. In fact many biologists indulge in vitalistic, in particular teleological expressions, as when they write about 'the purpose of organ X' or 'the use of process Y' or 'the plan (or design) of system Z'. To be sure they do not like being accused of vitalism, so they often prefer to use the term 'teleonomy' rather than 'teleology'. But this is just a verbal fig leaf attempting to hide the old Aristotelian final cause or purpose. So, if one is intent on collecting verbal evidence for teleological thinking among contemporary biologists, one is sure to come up with plenty of it. The question is to ascertain whether such a wealth of vitalistic phrases is a faithful indicator of the vitalistic nature of biology, or is just a relic of ancient biology or even of a nonscientific upbringing. This question cannot be answered by going over the same texts once again: it can be answered only by examining actual pieces of biological research.

Now, contemporary biology is observational, experimental, and theoretical. Since the concepts of vital force and purpose are theoretical, not empirical, it is useless to look for vitalism in biological observations or experiments. All such empirical operations can do is to supply evidence for or against the hypothesis that life is immaterial and that all life processes are goal-directed. The only place where such hypotheses might be found is theoretical biology. Let us therefore peek into it.

A number of branches of biology have been rendered theoretical in the modern sense, i.e. mathematical: population genetics (which embodies a good portion of the theory of evolution), physiology (in particular the study of biocontrol systems), ecology (in particular the study of competition and cooperation processes), and a few others. Every year hundreds of mathematical models of biosystems are published in the various journals of theoretical (or mathematical) biology. The author has been following this literature for two decades now without ever having seen a model — let alone an empirically confirmed model — that incorporates the hypothesis that life is an immaterial principle. Nor has he ever seen a mathematically correct and empirically successful model including the concept of goal directed process. What the recent literature shows us, on the other hand, is (a) an increase in the number of explanations of biological properties and processes with the help of physics and chemistry, and (b) an increase in the number of explanations of apparent purpose in terms of either control theory or the theory of evolution.

In conclusion, contemporary biology is not vitalistic even though many biologists do sometimes employ a vitalistic phraseology. (Remember that language is the clothing of ideas, and some clothes happen to disguise. Hence although philosophical analysis starts off with language it must go beyond it if it is to attain any depth and be of any use.) If anything, biology is becoming more and more materialistic in the process of studying living systems and their non-living components with the help of physics and chemistry — which does not mean that biology has been reduced to these sciences.

1.4. MIND IMMATERIAL?

Psychophysical dualism, or the thesis that there are minds alongside bodies, is probably the oldest philosophy of mind. It is part and parcel of most religions and was incorporated into philosophy by Plato. Descartes gave it a new twist by expelling all spirits from the body and turning the latter over to science — though retaining for theology and philosophy the rights over the soul. Many modern philosophers, as well as a number of scientists in their philosophical moments, have adopted dualism in some guise or other, some explicitly, most tacitly. Entire schools of thought have endorsed it, e.g. psychoanalysis with its talk of immaterial entities dwelling in the body, and anthropologists and historians with their talk of a spiritual superstructure riding on the material infrastructure. However, the fortunes of psychophysical dualism started to decline about three decades ago under the unconcerted action of philosophy and psychology. Let me explain.

There are at least three ways of undermining the doctrine of the immateriality of mind. One is to show that it is conceptually defective, another is to show that it is at variance with science, and a third is to exhibit a superior alternative. Let us sketch the first two tasks now, leaving the third for Chapter 5. (More on all three lines of attack in Bunge, 1980.)

The most blatant conceptual defect of psychophysical dualism is its imprecision: it does not state clearly what mind is because it offers neither a theory nor a definition of mind. All dualism gives us is examples of mental states or events: it does not tell us what is in such states or undergoes such changes — except of course the mind itself, so that it is circular.

A second fatal flaw of dualism is that it detaches mental states and events from any things that might be in such states or undergo such changes. This way of conceiving of states and events goes against the grain of science: indeed in every science states are states of material entities and events are changes of such states. Physiological psychology complies with this condition; psychophysical dualism does not.

A third grave defect of dualism is that it is consistent with creationism but not with evolutionism: indeed if mind is immaterial then it is above the vicissitudes of living matter, i.e. mutation and natural selection. On the other hand, according to materialism mind evolves alongside the brain. (See Chapter 6.)

But the worst feature of dualism is that it blocks research, for it has a ready answer to all the problems and it refuses to look into the brain to find out about mind. (Thus it enforces the separation between psychology and neurophysiology, and accordingly it favors verbal psychotherapy over behavior or drug psychotherapy.) By the same token dualism fosters superstition, in pàrticular belief in telepathy, psychokinesis, clairvoyance, precognition, and the various psychoanalytic immaterial entities — the ego, super-ego, id, and libido.

In short, psychophysical dualism is not a scientific theory or even a theory: it is just part and parcel of the old magical and religious world views: it is ideology not science. No wonder it is being replaced by the materialist approach according to which mind is a peculiar set of brain functions. More in Chapter 5.

1.5. CULTURE IMMATERIAL?

The idealistic philosophies of culture have accustomed us to think-ing of culture and cultural objects as immaterial. This view opens an abyss between man and other animals, as well as between the sciences of culture and all others. It also makes it hard to under-stand why the culture of a society depends upon, and coevolves along with, the economy and the polity of the society.

Historical and cultural materialists have criticized cultural ideal-ists and have tried to show that the material circumstances and activities of man — namely the natural environment, its transfor-mation through work, and the social relations deriving from this activity — determine everything else. (See Engels, 1878; Harris, 1979.) In particular, the intellectual and artistic culture, as well as the ideology of a society, become epiphenomena referred to collectively as the (ideal) "superstructure" mounted on the economic (material) "infrastructure". Thus historical and cultural materialism boil down, essentially, to economic determinism. To be sure this doctrine is often mellowed by adding that, once formed, the superstructure acquires a momentum of its own and can react upon the infrastructure. Still, the latter remains the first motor, and the superstructure is taken to be immaterial (or ideal) — a clear case of dualism.

I submit that historical and cultural materialism are only half way materialistic (because they include mind-matter dualism) and moreover cannot explain the actual interactions between the culture of a society and the other subsystems of the latter. That historical and cultural materialism are dualistic seems doubtless although it may not have been noted before: to a full-blooded materialist there is no such thing as an immaterial (or ideal) entity riding on a material one. And that the thesis of the absolute primacy of the economy over the rest is inadequate seems obvious when reflecting that a social change may be initiated either in the economy or the culture or the polity, and that some cultural changes — such as the introduction of literacy, reckoning, or science — have momentous economic and political effects.

An alternative view is this. A human society may be conceived of as a concrete (material) system composed of human beings. This system is in turn analyzable into four major subsystems composed of individuals: the biological, the economic, the cultural, and the political ones. The culture of a society, no matter how primitive it may be, is a system held together by information links, just as the biology is integrated by kinship and friendship relations, the economy is kept together by work and exchange links, and the polity by management and power relations. Hence the culture of a society may be regarded as a material system though not as a physical one, for it is characterized by nonphysical (emergent)

properties, such as creating and spreading knowledge, technical
know-how, and art.

A cultural activity is a brain activity of a certain kind, that influences the way other people think, feel, or act. The "product"
of such activity is called a 'cultural object', be it a poem or a
theorem, a cooking or medical recipe, a plan or a blueprint, a
sonata or a description of an animal, or what have you. As long
as such a "product" remains inside the skull of its creator, it is
only a brain process: it has got to be communicable to others in
order to rank as a cultural object. Such socializations or objec-
tifications need not be permanent but they must be accessible to
other people. A song that is never sung or written may be a thing
of beauty (for its creator) but cannot be a joy forever because it
cannot be transmitted and thus enjoyed by others.

To be sure we may feign, if we wish, that music and poetry,
mathematics and philosophy, biology and theology, are ideal (or
abstract) objects — provided we realize that they would not exist
were it not for their creators and users, all of whom are material
(though not physical) systems embedded in a social system. Even
the most complete library, museum, or laboratory in the world
would cease to be a cultural object after a nuclear holocaust, for
there would be nobody left to understand its contents. In other
words, World War III would leave no trace of Popper's "world 3".
Not because the nuclear blasts would destroy it — for only material
entities can be dismantled — but because there is no such "world 3"
to begin with. More in Chapter 8.

This materialist view of culture as a material system does not
debase or desecrate culture: it just demythifies it. On the other
hand the view that books, records, paintings and the like are in-
trinsically valuable, i.e. have an existence and a value of their own,
even in the absence of people capable of using them, is a crass
materialistic one, for it turns them into mere merchandise. (There
are of course deceptive cases. Thus a rock music record is nothing

but merchandise for, when played, it elicits no musical experience. Likewise many a book on esoteric matters, the reading of which brings neither understanding nor pleasure.) By avoiding reification and abstaining from placing values outside valuating brains, full-blooded materialism enhances the value of individual human beings, the sole known creators and consumers of cultural goods. Materialism becomes thus a strand of humanism.

In conclusion culture is not immaterial. If viewed as a process (of creation or diffusion), culture is just as material as motion or chemical change, for it occurs in and among ourselves, and we are material systems. And if viewed as a system — the system composed of producers and consumers of cultural objects — culture is a material thing. In either case culture is no less material than the economy or the polity. And it is not true that culture is always derivative or epiphenomenal: every important social event or process has a biological, an economic, a cultural, and a political component. Consequently it is not possible to develop a nation purely economically (or politically or culturally or biologically). Genuine development is at the same time biological, economic, cultural, and political: this is a corollary of our quadripartite conception of human society.

In sum, there is no good reason to suppose that culture is immaterial. On the other hand there are advantages — both intellectual and practical — to the thesis that the culture of a society is a material subsystem of the latter. We shall come back to this matter in Chapter 7.

1.6. CONCLUSIONS

It is now time to learn a couple of lessons from the above. One is that the concept of matter has changed over the centuries — or, rather, that there has been a historical sequence of concepts of matter. There is no reason to suppose that the present day concept of matter is final: after all, matter is what science studies, and as long as there is scientific research it is bound to come up with new concepts and new theories.

However, in order for a set of concepts to be called by a single name those concepts must share a meaning: otherwise we are in the presence of ambiguity, hence misunderstanding, not just of conceptual change. The historical sequence of concepts of matter complies with this condition: every member of it includes the idea that every material entity is changeable, at least with regard to place. To put it negatively: at no time has science asserted the absolute unchangeability or permanence of matter. We shall return to this in the next chapter.

Another lesson we can derive from the preceding sections is that, far from retreating from materialism, science is becoming more and more explicitly materialistic. It is doing so not only by shunning immaterial entities (life forces, ghosts, disembodied thoughts, supramaterial historical forces, etc.) but also, nay mainly, by studying material entities. Indeed science investigates physical things like quanta, fields, and bodies; chemical systems such as the organelles of cells; biosystems such as bacteria and fungi; and social systems such as economies and cultures. So much so that science may be characterized as the study of material things with the help of the scientific method and the aim of finding the laws of such things. In other words, scientific research presupposes and also enriches a materialist ontology. It behoves philosophers to unearth, develop, and systematize this ontology. Let us see next how this task can be accomplished.

CHAPTER 2

MATERIALISM TODAY

Materialism is a family of ontologies, or extremely general doctrines about the world. What all the members of that family have in common is the thesis that everything that exists really is material — or, stated negatively, that immaterial objects such as ideas have no existence independent of material things such as brains. Aside from this common core materialist ontologies may differ widely. It is only by adding further requirements that a definite materialist ontology will be individuated or built. We choose two: exactness and consistency with contemporary science. Let us look at these conditions.

2.1. EXACTNESS AND CONSISTENCY WITH CONTEMPORARY SCIENCE

So far materialism has been a rather amorphous body of somewhat vague beliefs. How can one transform such a doctrine into a system of clearly stated hypotheses consistent with contemporary knowledge, in particular logic, mathematics, natural science, social science, and technology? In general, how can one attempt to overhaul a philosophy? The short answer is: By substituting exact formulas for vague metaphors, by weeding out the obsolete theses, and incorporating new hypotheses consistent with contemporary knowledge.

Let us deal with exactification first. It consists in replacing vagueness with precision. This goal is attained by using, wherever necessary, the exact and rich languages of logic and mathematics instead of ordinary language, which is incurably fuzzy and poor.

(This *regula philosophandi* is probably Bertrand Russell's most important contribution to philosophy.) This condition suffices to disqualify dialectics — for being vague, unclear and metaphorical — as the worthy mate of materialism. Modern materialism is logical not dialectical. (More in Chapter 4.)

Here are a few examples of exactification at a very modest level of formalization.

EXAMPLE 1. "Events are changes in some material entity (i.e. there are no events in themselves)" is exactifiable as "For every event x there is one material object y and one change of state z of y such that $x = z$".

EXAMPLE 2. "Only material objects can act upon one another" transforms into "For any objects x and y, if x acts on y or conversely, then x is material and y is material'.

EXAMPLE 3. "Space is a mode of existence of matter" can be exactified as "Spatial relations are relations among material objects".

EXAMPLE 4. "Life is a form of matter" should be converted into "All organisms are material objects".

EXAMPLE 5. "A culture is a system whose living components are linked by information" may be exactified as "w is a culture if, and only if, there is an information flow x between every living component y of w and some other living component z of w".

The above formalizations employ only the most modest, albeit the most universal part of mathematics, namely ordinary logic. (For deeper reconstructions of ontological concepts and hypotheses, using more powerful formal tools, see Bunge, 1977a and 1979.)

Therefore they exhibit only the gross structure of the original statements. However, this often suffices to remove ambiguity or reduce vaguenesss. For example, the thesis "Change comes from opposition (ontic contradiction)" can be interpreted in several mutually incompatible ways. Two of them are "All change is generated by some opposition" (false), and "Some changes are generated by some oppositions" (trivially true). The whole of classical philosophy, in particular dialectics, is plagued by such ambiguities and obscurities. (See Chapter 4.)

Another merit of such exactifications is that they help to locate key concepts that should be elucidated in a second stage, such as those of material object, state, event, space, and life. Also, they show clearly that, whereas the first four constitute universal hypotheses, the fifth is a definition. Hence if we want our ontology to be scientific we must try and embed the first four statements into theories, and subject the latter to tests, whereas adopting the fifth is a matter of convention.

Almost any philosophy, provided it is not utterly irrationalistic like Heidegger's, or incurably absurd like Hegel's, can be rendered precise and clear, i.e. can be reformulated with the help of logical and mathematical concepts. (The apparent exception is ordinary language philosophy, which rejects this very move. But since linguistic philosophers do not profess to put forth any original substantive philosophical doctrines, they constitute no genuine exception.) Recall for instance the attempts of Whitehead, Russell, Carnap, and Goodman to turn phenomenalism into an exact philosophy. They met with success in the sense that their systems did constitute clear elucidations and systematizations of phenomenalism. But the results were shallow and barren as well as inconsistent with modern science, which is materialist and realist rather than phenomenalist.

Formalization then, though necessary for turning an unorganized body of vague theses into a hypothetico-deductive system, is insufficient for overhauling a philosophy. When we say that philosophy X is *obsolete* we intend to state that X fails to meet contemporary standards of exactness or that X is at variance with contemporary substantive knowledge about the world and human experience. Materialism is a case in point, for it not only inexact but has also failed to propose precise and up to date answers to the questions listed in our Preface. However, there is a difference between materialism and other philosophies, namely that its main tenets, however imprecise, are by and large countenanced by contemporary science. Indeed, as argued in Chapter 1, science investigates only material (or concrete) objects, and recognizes no immaterial ones – except for such objects as concepts, properties, and relations, none of which need be assumed to be self-existent.

So much for exactness as one of the two necessary conditions for the updating of materialism. Let us next apply the exactness rule and the condition of consistency with science to the definition of the concept of matter.

2.2. DEFINING MATTER

The most popular definitions of the concept of matter offered in the past are inadequate. Material entities cannot be identified with massive objects, let alone solid ones, since the discovery of massless fields such as the electromagnetic and neutrino ones. And material objects cannot be defined as those which exist independently of the subject, because an objective idealist will assert the autonomous existence of immaterial objects such as ideas. In short, whereas the first definition has turned out to be scientifically obsolete, the second has always been philosophically inadequate. (For further inadequate definitions see Cornman, 1971.)

We take our clue from contemporary science, according to which material objects, unlike ideal ones, are changeable. (Chapter 1.1.) Even the so-called elementary particles are either unstable or, if long-lived, they change in various ways by virtue of their interactions with other entities (particles or fields). On the other hand a conceptual object, such as the number 3 or the Pythagorean theorem, is not supposed to be in any state, let alone to undergo changes of state. Thus it makes no sense to ask 'How is 3 doing today?' or 'What is the equation of motion (or the field equation or the transmutation schema) of the Pythagorean theorem?'

We may then characterize a material object as one that can be in at least two different states, so that it can jump from one to the other. (Actually even the simplest material entity, such as an electron or a photon, can be at a given time in any of infinitely many different states.) That is, if x is a material object and $S_y(x)$ a state space for x, then the numerosity of the latter is at least 2, and conversely.

It might be objected that disembodied souls, such as were posited by Plato and Descartes, and the ghosts said to haunt the Scottish castles, are changeable yet immaterial, so they prove the inadequacy of our definition. Not so, for this definition happens to belong to a materialist ontology, where there is no room for disembodied objects, and where mental states are brain states. Besides, no state spaces can be built to represent immaterial objects; this is why mentalist psychology has remained nonmathematical.

(We need not go here into the technique for building a state space $S_y(x)$ for a thing x relative to a reference frame y, for which see Bunge, 1977a and 1979. Suffice it to say that it is a tacit epistemological postulate of contemporary science that, given any thing x of which we know some properties, it is possible (a) to represent each property of x by some mathematical function, and (b) to collect all such functions into a single function, called the *state function* of x, which (c) is assmed to satisfy some law statement. Each value of that function represents a state of x relative to the given reference frame y. The collection of such values, compatible with the laws of x, is called the *nomological state space* of x relative to y. As time goes by the thing moves from one state to another, slowly relative to some frames, fast relative to others.)

In short, we shall adopt

DEFINITION 1. An object x is a *material object* (or *entity*) if, and only if, for every reference frame y, if $S_y(x)$ is a state space for x, then $S_y(x)$ contains at least two elements. Otherwise x is an *immaterial object* (or *nonentity*).

More briefly,

$$\mu x =_{df} (y) \ (\text{If } S_y(x) \text{ is a state space for } x, \text{ then } |S_y(x)| \geqslant 2).$$

This definition allows one to partition every set of objects into entities and nonentities. It also allows one to construct

DEFINITION 2. *Matter* is (identical with) the set of all material objects.

In symbols,

$$M =_{df} \{ x | \mu x \}.$$

Note that this is a set and thus a concept not an entity: it is the collection of all past, present and future entities. (Or, if preferred, M is the extension of the predicate μ, read 'is material'.) Hence if we want to keep within materialism we cannot say that matter exists (except conceptually of course). We shall assume instead that individual material objects, and only they, exist. But this point calls for another section.

2.3. THE CENTRAL POSTULATE OF MATERIALISM

In order to state the central hypothesis of materialism we need not only the concept of matter but also that of reality for, according to materialism, all and only material objects are real. One way of defining the predicate "is real" is in terms of action or influence, namely thus. An object is real if, and only if, it influences, or is influenced by, another object, or is composed exclusively of real objects. (The second disjunct is needed to make room for the universe as a whole which, though uninfluenced by anything else, is itself composed of real entities.) More precisely, we propose

DEFINITION 3. An object x is *real* if, and only if, either (a) there is at least another object y whose states are (or would be) different in the absence of x, or (b) every component of x modifies the states of some other components of x.

DEFINITION 4. *Reality* is the set of real objects.

Note that, since "reality" has been defined as a set, it is itself unreal, for sets are incapable of influencing anything. (There is nothing wrong with this, for wholes need not possess all the properties of their parts.) Note also the contrast between Definition 4 and the vulgar or eclectic notion of reality as the sum total of all objects, whether or not they can act upon or be acted upon by other objects. Finally note that we are not defining "reality" as existence independent of the subject of knowledge, and this for two reasons. First because human creations do not come into being apart from us. (Thus a book, though real, owes its existence to its author and publisher.) Second because subjects too are supposed to be real.

We are now ready to state the hypothesis shared by all materialist ontologies:

POSTULATE 1. An object is real (or exists really) if, and only if, it is material. (Shorter: All and only material objects are real.)

This assumption bridges Definitions 1 and 3. By virtue of Definitions 2 and 4, Postulate 1 is equivalent to: *Reality is* (identical with) *matter*. Put negatively: Immaterial objects (nonentities) are unreal. In particular the properties, relations, and changes in either, of material objects are real only in a derivative manner: strictly speaking they are abstractions. For example, the distances among entities are not real: only spaced things are. Likewise events are not real: what is real is the entire changing thing. (However, there is no harm in talking about the properties of entities, and their changes, as being real provided we do not detach them from the things themselves.)

A standard objection to materialism can now be answered. It is that space and time, though surely immaterial, must be reckoned with: for, are not things supposed to exist *in* (regions of) space and time? The materialist answer is the relational theory (or rather theories) of space and time adumbrated by the preceding remark. According to it spacetime, far from existing on its own, is the basic network of changing objects, i.e. material ones. Hence instead of saying that material objects exist *in* space and time, we should say that space and time exist vicariously, namely by virtue of the existence (hence change) of material objects. Space and time do not exist independently any more than solidity or motion, life or mind, culture or history. (Cf. Bunge, 1977a.)

2.4. SYSTEM

We need next the notion of a system, which may be characterized as a complex object whose components are interrelated, as a consequence of which the system behaves in some respects as a unit or whole. Every system can be analyzed into its *composition* (or set of parts), *environment* (or set of objects other than the components and related to these), and *structure* (or set of relations, in particular connections and actions, among the components and

these and environmental items). It follows from the above definition of a system, jointly with Postulate 1 and Definition 3 that,
if a system is composed of material (real) objects, then it is real
itself. More precisely, we derive

THEOREM 1. A system is real (material) if, and only if, it is composed exclusively of real (material) parts.

This statement may appear to be trivial but it is not. For one
thing it tells us that systems other than physical or chemical systems, such as organisms and societies, are material. For another it
entails that, at least according to materialism, the "worlds" composed of ideas − such as fictions and theories − are unreal. What
are real are the creators of such ideal "worlds". More on this in
Chapters 7 and 8.

Now that we have the notion of a real (material) system, we
may as well add the assumption that renders materialism *systemic*,
to wit,

POSTULATE 2. Every real (material) object is either a system or
a component of a system.

Stated negatively: There are no stray things. The epistemological
consequence is obvious: Look for relations, in particular links (or
couplings or connections) among things.

Note the following points. First, our version of materialism is
dynamicist, for it identifies materiality with changeability. Given
the obscurities of dialectics, nothing would be gained and much
lost by appending the qualifier 'dialectical'. (See Chapters 3 and 4.)
Second, Postulate 1 should not be mistaken for nominalism (or
vulgar materialism, or reism), i.e. the thesis that there are only
things, properties being nothing but collections of things, and relations identical with tuples of things. True, we deny the indepen-

dent existence of properties and relations, but assert instead that things are propertied and related. Third, neither the postulate of materialism nor the accompanying definitions place any restriction on the kind of matter, i.e. on the composition of reality. In particular the above does not involve physicalism, or the thesis that every real object is physical. More on this in the next Section. Fourth, Postulate 2, or the hypothesis of systemicity, should not be mistaken for holism. Indeed, holism construes systems as wholes opaque to analysis. On the other hand we conceive of a system as a complex thing with a definite (though changeable) composition, environment, and structure — hence as analyzable.

2.5. EMERGENCE

Materialism is a kind of substance monism: it asserts that there is only one kind of substance, namely matter. (Substance pluralism, on the other hand, holds that there are multiple substances, e.g. matter and mind.) But materialism need not be property monistic, i.e. need not assert that all material objects have a single property, such as spatial extension, or energy, or the ability to join with other things. Materialism need not even assert that all the properties of material objects are of the same kind, e.g. physical. In particular, Postulate 1 and the accompanying definitions make room for property pluralism as well as the emergence hypothesis and conjectures about the level structure of reality.

Since the notions of emergence and of level are somewhat tricky and suspect in many quarters, we had better start by defining them. To this end we need a prior notion occurring in the definition of the concept of a material system, namely that of composition. The composition of a system is, of course, the set of its parts. More precisely, the A-composition $\mathscr{C}_A(x)$ of system x, or composition of x on level A, is the set of parts, all of kind A, of x. For example, the atomic composition of a molecule is the set of its atoms; the neuronal composition of a brain is the set of its neurons; and the individual composition of a social system is the set of persons composing it. We are now ready for

DEFINITION 5. Let x be a system with A-composition $\mathscr{C}_A(x)$, and let P be a property of x. Then
  (i) P is A-resultant (or resultant relative to level A) if, and only if, P is possessed by every A-component of x;
  (ii) otherwise, i.e. if P is possessed by no A-component of x, then P is A-emergent (or emergent relative to level A).

For example, the components of a cell are not alive: life is emergent, not resultant, relative to the components of a cell. And perception, feeling, and ideation are functions of multicellular neuronal systems, that no individual neuron can discharge: they too are emergent. On the other hand mass and energy are resultant properties.

There is nothing mysterious about emergence if conceived of in the above ontological sense. Emergence does become mysterious only when it is characterized epistemologically, namely as whatever property of a system cannot be explained from the components and their relations. But such characterization is incorrect, for one must then be able to state both the thesis of the explainability and that of the essential irrationality of emergence. (Needless to say I adopt only the former. But we shall not discuss it here because it is an epistemological thesis not an ontological one.)

Should anyone distrust emergence she should think of this. It is estimated that the going market price of the elementary components of the human body — i.e. carbon, nitrogen, calcium, iron, etc. — is about one dollar. On the other hand the market price of the biomolecules (DNA, RNA, proteins, etc.) of our body is about six million dollars. It has been said that this is the price of information; I would say that it is the price of emergent structure.

And now a hypothesis about emergence:

POSTULATE 3. Every system possesses at least one emergent property.

In a sense this hypothesis is trivial, for every system has a composition and a structure, which differ from those of its components. (Think of a system with three components held together by interactions of a single kind. It can be diagrammed as a triangle with the vertices representing the components and the sides the interactions. Remove now one component and compare the resulting system with the former.) However, the postulate is useful for it draws one's attention to emergence, a much misunderstood and maligned property and one whose recognition allows one to partition the family of materialist ontologies into two subsets. One is the class of ontologies that may be called *emergent materialism*, for they acknowledge emergence (e.g. Alexander, 1920; R. W. Sellars, 1922). Its complement is *physicalism* (or mechanism or reductive materialism), or the class of ontologies asserting that, whether on the surface or "at bottom" (or "in the last analysis"), every existent is physical (e.g. Smart, 1963).

2.6. LEVELS AND EVOLUTION

The emergence postulate suggests looking for emergence mechanisms, such as the clumping of like entities and the merger of unlike ones, as well as for developmental and evolutionary processes in the course of which systems of new species appear. At least the following comprehensive kinds or levels of entity may be distinguished:

> *Physical level*   = The set of all physical things.
> *Chemical level* = The set of all chemical systems
>   (wherein chemical reactions occur).

Biological level = The set of all cells or multicellular organisms.
Social level   = The set of all social systems.
Technical level = The set of all artifacts.

We cannot dwell on this taxonomy. Suffice it to note the following points. First, the components of every system belonging to a level above the physical one belong to lower levels. (Actually this relation serves to define the concept of a level in a rigorous manner: see Bunge, 1979.) Second, as we climb up the levels pyramid we gain some (emergent) properties but lose others. For example, the social level is composed of animals but is not an organism itself.

Finally we lay down a developmental hypothesis:

POSTULATE 4. The systems on every level have emerged in the course of some process of assembly of lower level entities.

Postulates 3 and 4 entail

THEOREM 2. Every assembly process is accompanied by the emergence of at least one property.

There is of course an enormous variety of assembly processes, from mere clumping to the merger of social systems, and there may be entirely new types of assembly processes in store. Moreover, while some of them have been natural others are man made: these are the cases of human social systems and artifacts (including living beings selected by man and, in the near future, synthesized).

In addition to such developmental assembly processes we make room for evolutionary processes, i.e. unique processes along which absolutely new things emerge, i.e. things possessing properties that no thing has possessed before. In biological evolution such novelties derive from mutation and adaptation; in cultural evolution, from behavior and ideation. Much more could be said about evolution on different levels, but we must close now by adding just one more assumption:

POSTULATE 5. Some processes are evolutionary.

Again, this postulate is far from trivial: creationism holds that all novelty is a gift of some deity, and physicalism (or mechanistic materialism) maintains that there is never novelty but merely rearrangement of pre-existing units. The above postulate is distinctive of *evolutionary* materialism.

2.7. CONCLUSIONS

The above postulates, theorems and definitions constitute the core of a research project aiming at building a new ontology. This ontology should be characterized by the joint possession of the following attributes:

(a) *exact:* every concept worth using is exact or exactifiable;

(b) *systematic:* every hypothesis or definition belongs to a hypothetico-deductive system;

(c) *scientific:* every hypothesis worth adopting is consistent with contemporary science − and therefore must stand or fall with the latter;

(d) *materialist:* every entity is material (concrete), and every ideal object is ultimately a process in some brain or a class of brain processes;

(e) *dynamicist:* every entity is changeable − to be is to become;

(f) *systemist:* every entity is a system or a component of some system;

(g) *emergentist:* every system possesses properties absent from its components;

(h) *evolutionist:* every emergence is a stage in some evolutionary process.

So much for the new ontology under construction. (For details see Bunge, 1977, 1979a, 1980.) Because the new ontology is supposed to possess all of the attributes listed above, it is hard to find a suitable name for it. 'Emergent materialism' would do no better than 'exact (or logical) materialism'. However, a name is needed for practical purposes. If pressed to choose we should pick the most comprehensive. This one seems to be *scientific materialism*, for in our century "scientific" embraces "exact", "systematic", "dynamicist", "systemist", "emergentist", and "evolutionist". This explains the title of the present book.

To be sure there is some overlap between scientific materialism and alternative materialist philosophies — otherwise it would not deserve bearing the family name 'materialism'. However, all other materialist ontologies lack at least one of the characteristics listed above. In particular most of the alternatives are inexact (verbal or metaphorical rather than mathematical and literal); they are unsystematic or fragmentary, or they are dogmatic (unchanging and therefore quickly dated) rather than keeping abreast of science; or they are atomistic rather than systemic, or physicalistic (mechanistic) rather than emergentist. In the case of dialectical materialism — the most encompassing, popular and influential philosophy today — dogmatism comes from partisanship, and inexactness from dialectics. However, the obscurities of dialectics call for another chapter.

PART TWO

BECOMING

CHAPTER 3

MODES OF BECOMING

Science and philosophy study what there is and how what is changes into something else: science does it in detail, philosophy in outline. The study of being goes hand in hand with the study of becoming. Thus we cannot know what an elementary particle is without finding out its modes of generation and transformation. Nor can we know what man is without inquiring into the mechanisms of human conception and development, evolution and extinction. In every department of science the focus is on transformation, but we could not even describe a transformation without having some idea of what it is that changes. So, the study of being and the study of becoming are two sides of one and the same investigation.

We may group the objects of scientific study either as to their mode of being or as to their mode of becoming: no matter what we may start with we shall be led to the complementary aspect. Thus we may class existents into physical things, chemical systems, biosystems, minding systems, societies, and artifacts. Or we may class modes of becoming into chaotic, random, causal, synergic, conflictive, and purposeful. Focusing on a single mode of being or becoming at the expense of all others gives rise to a particular ontology – a onesided view of the world to be sure. Only an integration of the various possible modes of being and becoming will yield a realistic ontology, i.e. one compatible with our scientific knowledge of reality. See Table I and Figure 1.

It is doubtful whether radical indeterminism, or the denial of regularities of all kinds, has ever been held consistently. To be sure Epicurus endowed his atoms with a spontaneous swerving motion

TABLE I
Basic modes of being and becoming, and their respective ontologies

Mode of being	Ontology	Mode of becoming	Ontology	
Physical	*Physicalism*	Chaos	*Indeterminism*	
Chemical	*Chemism*	Randomness	*Probabilism*	⎫
Biological	*Vitalism*	Causation	*Causalism*	⎪
Psychical	*Animism*	Synergy (cooperation)	*Synergism*	⎬ *Determinism*
Social	*Sociologism*	Conflict (competition)	*Dialectics*	⎪
Technical	*Machinism*	Purpose	*Teleology*	⎭

(*clinamen*), or irregular deviation from rectilinear motion; however, such deviations were thought to be rather small. Also, one century ago Emile Boutroux and Charles S. Peirce wrote about deviations from laws, but they seem to have had in mind inaccuracies in our representations of regularities, as well as measurement errors, rather than lawlessness. In sum radical indeterminism, or the total denial of lawfulness, has found no favor with philosophers. No wonder, for philosophy is a search for pattern — just like science, only more general.

Quantum mechanics has occasionally been regarded as endorsing radical indeterminism, but mistakenly so, for every scientific theory centers around a set of law statements, and quantum mechanics is no exception. And some applied mathematicians and theoretical biologists speak occasionally about "chaotic" solutions to certain nonlinear differential equations. But since such equations are deemed to represent regularities or laws, the word 'chaos' is misplaced and should be replaced by 'non-periodic' or at most 'mock-random'.

In sum, science has no use for chaos or lawlessness, hence it does not condone radical indeterminism. Nevertheless it would be foolish to deny that there are accidents on all levels, and in particular that human life is a tissue of accident and necessity. But such accidents are crossings of lawful lines.

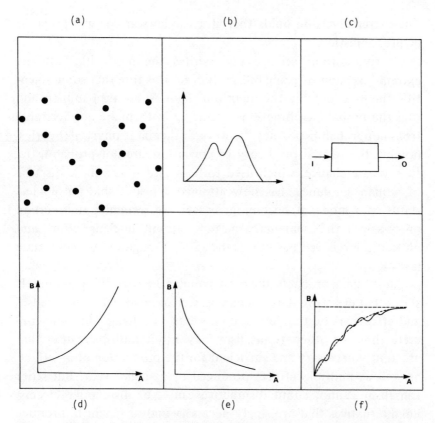

Fig. 1. Models of modes of becoming. (a) A chaotic distribution: no law. (b) A probability distribution (stochastic law). (c) A causal black box: transforms inputs into outputs (causal law). (d) Cooperation: A helps B grow and conversely. (e) Competition: A inhibits the growth of B and conversely. (f) Purposefulness: Process approaches goal despite local setbacks.

As for probabilism, or the thesis that there is objective chance or randomness, albeit always lawful, it is a rather modern idea that does not go farther back than Antoine-Augustin Cournot. There are two versions of probabilism, a moderate and an extreme one. The moderate version holds that there are primary (irreducible, basic) probability laws, such as the quantum-mechanical ones.

The extreme version holds that all basic laws are or will prove to be probabilistic.

Clearly, current science underwrites the moderate, not the extreme, version of probabilism. Indeed, it is true that some scientific theories, notably the quantum theories, are probabilistic and that the probability functions occurring in them are not derivable from nonprobabilistic ones. However, other basic physical theories, notably the relativistic theory of graviation, are not probabilistic. In any case chance, which used to be regarded as a mere disguise of human ignorance, has now attained a respectable ontological status as a mode of becoming: indeed we recognize that certain processes at the elementary particle, atomic nucleus, atom, and molecule levels are basically random – though at the same time lawful.

Causalism is probably the most popular doctrine of becoming. It states that every event has a cause as well as an effect, where causes and effects are best understood as events, i.e. changes in some concrete thing or other. (Causalism may admit multiple causes that are jointly necessary and sufficient for the production of an effect, as well as multiple effects produced by a single event. But strict causalism cannot admit disjunctive causes or disjunctive effects, i.e. alternatives that are singly necessary and sufficient to produce an effect, or alternative effects of a single cause, because either would open the door to probabilism.)

It is often stated that the quantum theory has refuted the causal principle. In my view this theory has only restricted the causal domain. Moreover the theory has a causal strain, manifest in the way it treats the probability that a given cause (e.g. a field force) will produce a certain effect (e.g. the scattering of a particle within a given angle). That is, quantum mechanics and quantum electrodynamics have both a stochastic and a causal aspect, so they call for an ontology allowing for the coexistence and intertwining of both categories. (See Bohm, 1957; Born, 1949; Bunge, 1959; Cassirer, 1956.)

Reciprocal causation, or interaction, is far more common than either pure randomness or one sided causation. There are of course many kinds of interaction. While some have only quantitative effects, others produce qualitative changes. The most interesting among the latter are the synergic and the conflictive interactions. The former, i.e. cooperation, leads to the formation or maintenance of systems of all kinds — physical, chemical, biological, and social. On the other hand conflictive interaction may end up in the destruction of some or even all of the entities in conflict, as exemplified by matter-antimatter collision, predation, and natural selection.

Heraclitus emphasized conflict at the expense of cooperation, and initiated a whole class of dialectical ontologies, each confirmed by a number of examples and refuted by others. I believe it is mistaken to extol one form of interaction at the expense of others. In nature and in society we see or conjecture cooperation as well as conflict, so our theories of qualitative change should make room for both. Before we attempt to explain the disintegration of a system as an outcome of inner conflicts we must be able to account for its emergence as a result of cooperation — mostly unwitting of course.

Finally teleology, or the doctrine that whatever happens is directed towards some goal, is perhaps the oldest of all world views. We find it in primitive ideologies, particularly in religions, and in prescientific philosophy. There are two kinds of purposiveness: transcendent or other-directed, and immanent or inner-directed. The religious world views uphold of course transcendent teleology, whereas thinkers such as Aristotle and Lamarck, who believed in the purposive nature of biological processes but were at the same time naturalists, favored immanent teleology.

Both forms of teleology died with the birth of modern science: nowadays they are found, if at all, in the ideology of some scientists rather than in the product of their research. Indeed, the concept of purpose does not occur in either the theories or the data of physicists, chemists, or biologists. Purpose has been explained (or replaced) either by control (or negative feedback) or by genic variation followed by selection (or elimination of the maladapted.)

As for psychology, only psychoanalysts and parapsychologists insist that all mental phenomena — even dreams and neuroses — serve some purpose, such as ego protection, wish fulfillment, or anxiety avoidance. Scientific psychology is nonteleological. Yet it does not deny that higher vertebrates can behave with goals in view (or smell or taste). Only, instead of attempting to explain behavior in terms of an irreducible immaterial purpose, scientific psychologists try to account for purposive behavior in terms of neurophysiological processes stimulated and constrained by genetic and environmental determinants. In sum, teleology is dead *de jure* though not *de facto*, and some scientists are busy trying to explain purposive behavior in nonteleological terms.

To summarize. Contemporary science recognizes five main modes of becoming: randomness or lawful chance, causation, co-operation, conflict, and purpose. The first four seem to operate on all levels, whereas purposive behavior seems to operate exclusively among higher vertebrates. So, science cannot be said to countenance either radical indeterminism, or probabilism, or causalism, or synergism, or dialectics, or teleology.

Instead, contemporary science seems to adopt an eclectic or integrative stand with regard to basic modes of becoming. Or, if preferred, science seems to favor an ontology making room for all five categories of becoming. In particular, such an ontology inspired by contemporary science will tend to view man as a bio-psychosocial system engaged in processes where randomness and causation intertwine with cooperation, competition and purpose. A concentration on either of these modes of becoming with ignorance of the other four results in a distorted picture of reality and therefore in a poor guide to intelligent and effective action. Dialectics is a case in point and will be tackled next.

CHAPTER 4

A CRITIQUE OF DIALECTICS

A central claim of this book is that, whereas materialism is true albeit underdeveloped, dialectics is fuzzy and remote from science. So, if materialism is to develop along the lines of exactness and in harmony with science, it must keep clear from dialectics. Let me substantiate these charges against dialectics.

4.1. THE PRINCIPLES OF DIALECTICAL ONTOLOGY

We shall be concerned here with dialectical ontology. I submit that dialectical ontology has a plausible kernel surrounded by a mystical fog. The plausible kernel of dialectics is constituted by the hypotheses (i) that every thing is in some process of change or other, and (ii) that at certain points in any process new qualities emerge. However, this nucleus (a) is common to all process metaphysics and (b) it has yet to be expanded into a general, exact and consistent theory before it can be claimed to be in fact such a theory.

As for the fog surrounding the plausible kernel of dialectics, it consists mainly in the theses (iii) that to every object there is an antiobject, (iv) that all opposites fight each other, this conflict resulting either in the annihilation of one of them or in some new object synthesizing both contradictories, and (v) that every stage in a development negates the previous stage and furthermore that two successive negations of this kind ensue in a stage similar to but also somehow superior to the original stage.

It will be shown that the main source of confusion and obscurity are the ambiguous key expressions 'dialectical negation' and 'dialectical opposition'. The removal of some of this ambiguity will leave us with an intelligible doctrine. However, this weaker dialectics can claim no universality. And, because it still focuses on a very special relation — that of opposition — it is at best a special or limit case of a far richer theory of change conceived in the spirit of science rather than of pre-Socratic philosophy.

Dialectics is supposed to be so well known that few if any contemporary dialecticians take the trouble of stating its theses with precision, in any detail and systematically, or even clearly. This lack of precision, detail and systemicity gives rise to more commentaries than original works and explains why there is such a bewildering profusion of interpretations of dialectics. We must therefore start by trying to dig up the principles of dialectics or at least of our version of it. (Should anyone feel dissatisfied with this version he is welcome to produce a more satisfactory formulation. In fact it is high time that somebody did it.)

We take the principles of dialectics to be the following (Hegel, 1816, 1830; Engels, 1878, 1872–82; Lenin, 1914–16; Stiehler, 1967; Pawelzeig, 1970; Narski, 1973; Bunge, 1973a).

D1   Everything has an opposite.

D2   Every object is inherently contradictory, i.e. constituted by mutually opposing components and aspects.

D3   Every change is the outcome of the tension or struggle of opposites, whether within the system in question or among different systems.

D4   Development is a helix every level of which contains, and at the same time negates, the previous rung.

D5   Every quantitative change ends up in some qualitative change and every new quality has its own new mode of quantitative change.

We regard the first three principles as peculiar to dialectics.

However, we shall examine all five axioms and will attempt to reformulate them in a clearer way, so as to be able to evaluate them.

4.2. THE THESIS THAT FOR EVERY THING THERE IS AN ANTITHING

The dialectical thesis *D1*, that to every object there corresponds an antiobject, is ambiguous both because of the ambiguity of 'object' and that of 'anti'. In fact it splits into at least two propositions:

D1a   For every thing (concrete object) there is an antithing.

D1b   For every property of concrete objects there is an antiproperty.

(There are further possibilities, involving conceptual objects as well as anticircumstances and antievents. We discard them assuming that dialectical ontology is concerned with concrete objects and that both circumstances and events can somehow be reduced to things and their properties.)

These two hypotheses are still unclear as long as it is not explained what antithings and antiproperties could be. Let us attempt to remove this unclarity, starting with the notion of an antithing. There are at least four construals of the term 'antithing' (or 'dialectical opposite of a thing'):

(i) The antithing of a given thing is the absence of the latter (e.g. antilight is darkness). But the absence of some thing cannot oppose the latter, let alone combine with it to form a third entity. Hence this definition is inadequate: the dialectical opposite of a concrete thing cannot be nothingness.

(ii) The antithing of a given thing is the environment of the latter, i.e. its complement in the totality of things or universe. This definition too is wanting, because there need be no opposition or strife among complementary things: just think of our solar system and the rest of the world.

(iii) An antithing of a given thing is an entity such that, if combined with the latter, destroys it in some respect and to some extent, as when water extinguishes fire or a poison kills a plant. If this definition is adopted then one cannot guarantee the existence of an antithing of any given thing. And in the cases where there are antithings they may not be unique: there are many fire extinguishers besides water, and for every weed there are several weed killers. Hence this definition too is unsuitable.

(iv) An antithing of a given thing is an entity such that, when combined with it, produces a third thing that somehow both contains and supersedes the two. At first blush a particle and an antiparticle (e.g. a proton-antiproton pair) constitute such a pair of dialectical opposites. But as a matter of fact they do not, for they may fuse into a photon, which does not sublate the original entities but is a thing of an altogether different kind. Another pseudo-example is polymerization: this is a synthesis all right but one of equal rather than opposite entities.

None of the above four definitions of an antithing seems to serve the purposes of dialectics. Consequently either *D1a* makes no sense or a fifth, more suitable definition of antithing is needed. If the former is the case, nothing more need be said. If the latter, then it behoves the dialectician to contrive such a redefinition — or to grant that *D1* does not concern things and antithings. If he resorts to the help of Aristotle's *Categoriae* he will learn that dialectical opposition concerns traits or properties (actual or potential) rather than things. Let us then explore this other possibility.

4.3. THE THESIS THAT FOR EVERY PROPERTY THERE IS AN ANTIPROPERTY

We turn now to thesis *D1b* in Section 4.2, namely that to every property there corresponds an antiproperty. This sentence will make hardly any sense unless the term 'antiproperty' is rendered significant. Here again there are various possible interpretations. We shall consider the following four:

(i) The antiproperty of a given property is the absence of the latter, as in the case of good and not-good (which is bad or neutral). Thus if a predicate P represents a given positive property, such as that of being wet, or of interacting, then its negate *not-P* would represent the corresponding antiproperty. However, a property and its absence cannot combine to produce a third trait, namely the synthesis of the former, simply because the absence of a given characteristic is not a trait effectively possessed by a thing. Negating P (or asserting that a certain object satisfies the predicate *not-P*) is a strictly conceptual operation without an ontic counterpart. And conjoining P with *not-P* yields the contradictory or null property, i.e. that which no object, whether conceptual or material, possesses. We must therefore reject the proposed identification of anti-P with *not-P*. (More on negative properties at the end of this Section).

(ii) The antiproperty of a given property is the complement of the property in the set of all properties. This definition too is wanting, because an individual property is not on the same footing as a collection of properties, hence cannot oppose it, let alone form a third property synthesizing the two.

(iii) An antiproperty of a given property is a property that can check, balance or neutralize it, as when pushing and pulling compensate one another and keep a thing stationary. This construal makes sense, and one can find instances of antiproperties of this kind. The trouble with it is that antiproperties of this kind are neither universal nor unique. That is, not every property has an antiproperty and, when a property does happen to have an opposite number, the latter may not be unique. For example, the property of having a mass has no opposite in this sense, for there is no

such thing as antimass or negative mass. And the growth property
can be checked by a number of opposing properties. In short this
construal, too, falls short of the needs of dialectical ontology. But
at least it makes sense.

(iv) An antiproperty of a given property is a trait such that,
when conjoined with the property in question, gives rise to a third
property that subsumes the two and is not null. The combination
of an acid with a base to yield a salt would seem to exemplify this
sense of property opposition. But it may also be regarded as a
combination of opposite things. Furthermore, although there are
examples there are also counterexamples. Thus the mere accretion
of like particles (with no opposition whatever) gives rise to massive
bodies. Moreover such a process may reach the point of collapse —
a qualitative jump involving no dialectical opposition. In sum the
fourth construal of 'antiproperty', though significant, does not
justify the prefix 'anti' and does not give rise to any universal law.

The upshot is this. Of the four plausible interpretations of 'antiproperty' we have considered, two (i.e. (iii) and (iv)) make good
sense but neither allows one to assert *D1b* in all generality. Only a
much weaker thesis can be asserted, namely

D1c   For some properties there are others (called their 'anti-
          properties') that counteract or neutralize the former.

In simpler terms: Some things oppose others in certain respects.
But this is a triviality, so no dialectician should be willing to settle
for it. If he does not then he must supply a fifth, more suitable
definition of 'antiproperty'. However, even assuming that he solves
this problem, he will have to face the following difficulty.

The assumption that every property is matched by an antiproperty is possible in an idealistic ontology that refuses to make a
sharp distinction between a predicate (a concept of a certain kind)
and a property of a concrete thing, such as that of being extended.
And, because the concept (predicate) *not-P* is just as legitimate as

the concept P, to a Platonist and to a Hegelian a negative property will be as real as a positive one. The dialectical idealist will be able to admit thesis $D1b$ provided he manages to produce a suitable elucidation of the notion of an antiproperty.

Not so the dialectical materialist if he takes materialism seriously. In fact to a nonidealist things have only positive properties: although there are negative predicates, these cannot represent any properties of concrete objects. To him, if a predicate P represents a certain property, then its negate *not-P* does not represent an antiproperty but just the absence of the property represented by P. In fact, if the formula "Pa" abbreviates the statement "Thing a has property P", then the formula "$not-Pa$" abbreviates "Thing a does not have property P" or, in the alethic interpretation, "It is false that thing a has property P". Since the absence of a trait cannot be said to be the dialectical opposite of that trait, it follows that negative predicates fail to represent antiproperties. Negation is a conceptual operation without an ontic counterpart: it has to do with statements and their denials, not with the struggle of ontic opposites. (For similar statements see Hartmann, 1957; and Kraft, 1970.)

Likewise the disjunction of predicates does not represent disjunctive or alternative properties. There is no such thing as a man with two or three legs, even though the proposition "Men have two or three legs" is true. Disjunction is as conceptual and anontic as negation. (For details see Bunge, 1977.) This will have an important consequence for the thesis that formal logic is a limiting case of dialectics (see Section 4.9). And it has an equally disastrous consequence for the claim that all knowledge is a picture of reality. In fact consider the set of predicates of a given order (arity) and a common reference, such as the totality of unary predicates concerning mammals. This predicate set is a Boolean algebra. On the other hand the corresponding set of properties of the same individuals (of mammals in the example) is just a semigroup where the concatenation is interpreted as the conjunction of properties. To claim that the structure of predicates mirrors the structure of properties is consistent with an idealistic ontology but inconsistent with naturalism, in particular materialism, which has no use for negative properties. Dialectical idealism is then possible though implausible, whereas dialectical materialism is implausible and inconsistent with the reflection theory of knowledge.

4.4. THE THESIS THAT EVERY THING IS A UNITY OF OPPOSITES

The thesis *D2*, that every object is a unity of opposites, is usually regarded as constituting the essential thesis of dialectics. But, again, the sentence makes hardly any sense unless the term 'opposite' is rendered precise. And this, as we saw in the last two sections, is no easy task and in any case it has not been performed by the dialectical philosophers.

I submit that *D2* does make sense if opposition, or ontic contradiction, is construed as a relation between properties, namely that of counteraction or neutralization (sense (iii) in Section 4.3). We adopt then the following definition: "Property (or relation) P_1 is said to oppose property (or relation) P_2 if, and only if, P_1 tends to check (neutralize, balance, or dim) P_2 and conversely". For example, in an overpopulated country population growth and welfare are mutually opposite because the former defeats the attempt to keep and raise the standard of living.

If opposition is taken in this sense then one can assert that there are systems ridden with internal ontic contradictions. But this is a far cry from claiming that all systems are contradictory. For example, according to contemporary physics electrons and photons have no inner contradictions. Which is just as well, for if every thing were composed of mutually contradictory parts then every such part would be composed similarly and one would be faced with an infinite regress.

Now if all we can say is that *some* things (or parts thereof) oppose others in *some* respects (which was our thesis *D1c* in Section 4.3) then all we can conclude is that *some* systems have components or traits that oppose each other in *some* respects. That is, we get the weaker thesis:

D2a   Some systems have components that oppose one another in some respects.

The central thesis of the unity of opposites remains thus restricted to complex things and moreover to some aspects of them. Nothing is said about simple things (if there are any); nor is anything said about every aspect or property of any system. The watered down version *D2a* of the central thesis of dialectics is not universal, hence it cannot be part of a sweeping theory of change. Moreover, what it suggests, namely the analysis of systems into poles, constitutes no advance in ontology. Rather on the contrary, thinking in opposites is characteristic of the archaic mentality (see Frankfurt *et al.*, 1946) as well as of Greek classical thought with the significant exception of the atomists (see Lloyd, 1966). Which is not surprising, since it is simplistic, and every early mode of thinking is simplistic.

That thinking in opposites involves a brutal oversimplification of the real world can be clearly seen in the following example. A system may be said to be a *polar system* if it is composed of parts that can be in either of two states, such as closed and open, up and down, or active and inactive, which are mutually exclusive or contradictory in a strong sense. A switching circuit, as employed in a digital computer, may be regarded as a polar system. But obviously this is an oversimplification, since it accounts only for the net outcome of a process, neglecting the intermediate states or transients. If these transient states are disregarded then the operations of the system can be described with the help of ordinary Boolean algebra. But if we wish to take the transient states into account then the state space acquires a third member and it is endowed with a richer structure, namely a three valued Łukasiewicz algebra (cf. Moisil, 1971). And even a three state machine is an oversimplification of interest to the computer designer and user but of little interest to the physicist. Indeed, lumping a whole continuum of steady states into the open and closed categories, and likewise the whole continuum of transients into one, is just a first and very coarse approximation. A more truthful account is given by electrical network theory, and an even deeper description by electrodynamics, both of which assume infinitely many states, whether steady or transient. In this true account there is no trace of polarity. Polarity resides in our thinking about reality rather than in the world itself; moreover polarity is a mark of incipient knowledge, not of science.

4.5. THE DIALECTICAL VIEW OF CHANGE

That some changes result from strife or conflict of some kind or other is rather obvious. Conspicuous examples are of course competition among animals and war among humans. It would be foolish to deny them. What is being questioned is whether competition is universal to the point that it is behind every change. And it seems equally obvious that this is not the case, i.e. that there are changes which are not brought about by any ontic contradiction. For example, the motion of a particle or of an electromagnetic wave in free space are not conflictive. Nor is the formation of a hydrogen molecule out of two hydrogen atoms, if only because the latter cannot be said to oppose one another; quite on the contrary, they might be said to cooperate.

The most we can accept is the weaker thesis:

> *D3a*   Some changes are brought about by the opposition (in some respects) of different things or different components of one and the same thing.

But this is almost trivial. Any theory of competition (e.g. chemical kinetics, Volterra's theory of the growth of competing species, and game theory) is far more precise and rich than that.

Moreover a literal interpretation of the dialectical principle of contradiction as the motor of change is inconsistent with the reflection theory of knowledge. Indeed, if every statement reflects something real, then every contradictory statement must reflect some ontic contradiction, which is in turn the source of some change. But since a contradiction is false, it cannot reflect anything real. Hence either there is no change or the reflection theory of knowledge must part company with dialectics. Which is the conclusion we reached at the end of Section 4.3.

It is sometimes held that, although dialectics may be false of nature, it must hold for society: witness the permanent social and international tensions at all levels, which from time to time erupt as open wars. This belief is false because it ignores cooperation, which is more basic, conspicuous and desirable than conflict. Every social system, from family to community to nation to world system, emerges and subsists by cooperation, whether deliberate or unwitting. Surely tension is rampant, but the whole point of certain institutions is to prevent tension from disintegrating society, whereas that of others is to facilitate cooperation. Think of industry, trade, and education, all of which necessitate organized cooperation. Even large scale conflict, whether economic or political, would be impossible without cooperation. Therefore to single out conflict as the sole source of change is to refuse to see the other side of the coin and therefore to remain ignorant of reality. Thus dialectics is no less an effective obstacle to the understanding of society than synergism.

Could it be that dialectics, though not the overriding pattern of natural and social processes, is that of the development of thought and, in general, of human knowledge? This restriction of the reach of dialectics has been advocated by a number of philosophers from

Plato to Bachelard (1940) and Popper (1940). In particular, the latter claims to have generalized the dialectic of knowledge, keeping its thesis that contradiction (in particular criticism) is the motor of the growth of knowledge (which he in turn identifies with the development of thought), particularly in the sciences. But surely somebody must have thought up a thesis, in response to some problem, before its antithesis could be advanced: creation precedes destruction. What criticism effects is not to replace creation but to keep it under control. The sheer clash and discussion of opinions – typical of theology, political ideology, and too much of philosophy – is no substitute for problem finding, model building, theorem proving, and experimenting.

Besides, in a materialist perspective there is no development of knowledge in itself simply because there is no knowledge in itself ("objective knowledge", or knowledge without a knowing subject). What is real is the cognitive development – or stagnation or retrogression – of human brains and communities of researchers and knowledge users. So, if we wish to discover the real motor(s) of cognitive development we must look at the cognitive activities of real knowing subjects (individuals or teams) embedded in their societies, rather than at the disembodied products of such activities. (And we are likely to find two such forces rather than one: sheer animal curiosity and need – or, if preferred, cognitive and practical problems.) And since cooperation – the sharing of problems, theories, methods, etc. – plays a central role in such cognitive activities, the "development of knowledge" does not fit the dialectical "law" of change. Whether it fits any laws at all, aside from the psychological ones, we do not know.

Having deprived *D3* of universal generality let us now try to understand how it may have come about. (Caution: This piece of archaeology of ignorance is purely conjectural. But at least it does not fit a preconceived schema such as the dialectical triad.) To this end it will be convenient to reformulate *D3* to read: "There is change in system x just in case x has an internal contradiction, or there exists another system y such that x and y oppose each other in some respect". This statement can be obtained by hasty generalization and logical fallacy, namely thus. One starts by looking around and notes that, wherever there is ontic "contradiction" (internal tension or strife), there is change. One then jumps (invalidly) to the conclusion that the converse holds as well, i.e. that strife is "father of all and king of all" (Heraclitus, Fragment 53). From these statements the given "law" follows. Finally one looks for confirmers of the thesis. And sure enough one finds them, particularly if one adopts each time a convenient sense of 'contradiction' or 'opposition'. In the process one discards all counterexamples. And so one remains at the stage of uncritical black-and-white thinking peculiar to archaic eras.

4.6. THE DIALECTICAL VIEW OF DEVELOPMENT

Consider now the dialectical thesis $D4$ of the helicoidal ("spiral") nature of all development, whether in nature, society, or thought. This thesis, too, is unclear because of the uncertainty about the expression 'dialectical negation' in this context. What we usually get from dialecticians by way of clarification is a few alleged examples, such as the plant being "negated" by its seed, which by germinating and developing into new plants "negate" themselves. This kind of negation – to keep the archaic terminology – is often called *Aufhebung* (sublation), to mark its difference from the concepts of negation involved in the other theses of dialectics. We are also told that the double negation of x is usually superior to x, except perhaps in the case of mathematics. This is all we learn from the classical writings on dialectics, from Hegel to Mao.

There are several difficulties with the "law" $D4$. For one thing, if the dialectical negate $-x$ of x succeeds x, then the two cannot combine to form a third object, because the sublation of x involves

the demise of x. Secondly, the thesis asserts that every development is progressive: in fact it says that $-x$ is superior to x. But
this dogma ignores stasis and decline. Thirdly and most important,
because the concept of sublation is foggy, so is $D4$. And being a
misty sentence we can hardly pass judgment about its truth value.
(An imprecise sentence designates ambiguously not just one proposition but a multitude of statements and moreover an indeterminate collection of such, i.e. one with no clearly identifiable
members.) While waiting for the dialecticians to clarify the notion
of *Aufhebung* and reformulate $D4$ in a clear way, we should by-
pass these stagnant waters and proceed to formulating clear, cogent,
and general theories of developmental and evolutionary processes
in harmony with science. This should prove more rewarding than
attempting to force reality, as well as the rich scientific theories,
into a preconceived and simple-minded polar pattern.

It might seem, though, that catastrophe theory − so fashionable
these days − has given dialectics a new lease on life and has even
conferred mathematical respectability upon it. Indeed the very
creator of the theory has explicitly stated that "Our models
attribute all morphogenesis to conflict, a struggle between two or
more attractors" (Thom, 1975; p. 323). But the Heraclitean aura
of catastrophe theory is illusory, for the conflicting objects are
geometrical, not material. Besides, the catastrophe-theoretic models
in biology and sociology are descriptive and nonpredictive, whereas the point of dialectics is to explain change and expedite it.
Moreover, such models cannot be taken too seriously because they
are inconsistent with a number of established theories − in particular genetics and the theory of evolution −and they ignore all
stochastic features. Worse, they are shot through with arbitrariness
and even mathematical sloppiness (Sussmann, 1975). To top it all,
the ontology of catastrophe theorists is Platonic ("Form rules over
matter and pre-exists it"), and so it can hardly jibe with materialism. In conclusion, dialecticians can expect no comfort from
catastrophe theory.

4.7. THE QUANTITY-QUALITY THESIS

D5 is perhaps the most popular of all the theses of dialectics. It is also the one that has been formulated in the most ludicrous way, namely as the law of the *conversion* of quality into quantity and vice versa. This is of course unintelligible. A quantity is either the numerosity of a set of things, such as the population of a town, or the numerical value of some quantitative property, such as the probability of a given transition from one state to another state of some thing. In any case quantity does not oppose quality if the latter is taken to be identical with property.

Certainly "quantitative" and "qualitative" are mutually contrary in the strict or formal sense that the second can be defined in terms of the former and of logical negation, namely thus: "If *P* is a property, then: *P* is *qualitative* $=_{df}$ *P* is not quantitative". But this is not the sense in which *D5* opposes quantity to quality. Indeed it would be blatantly false to say that a qualitative property, such as separation, gets transformed into the quantitative property of distance, and conversely, as part of a lawful natural process. In our construal of *D5* all it asserts is that in every process a stage comes when some new property emerges, which has in turn its own mode of quantitative variation. Thus urbanization leads to cities, not to large villages, and once a city is constituted it grows or decays in a peculiar way — its dynamics differs from that of the village.

Stated as *D5* the quantity-quality thesis makes good sense and it may even be true. Nevertheless it should be entertained as a hypothesis rather than as an article of faith, as should indeed every other ontological principle. Moreover it would be worthwhile to try and obtain *D5* as a theorem in a general theory of change.

4.8. THE UNIVERSALITY CLAIM

All five theses of dialectics are alleged to be universally true. However, our analysis has shown that, to the extent to which $D1$, $D2$, and $D3$ make sense, they should be replaced by statements of restricted scope. (Thus, given that some novelty-producing processes consist in the clumping of equals rather than in the collision of opposites, $D3$ is false and its denial is true.) If $D4$ should become meaningful, the chances are that it, too, is an existential, not a universal, statement. Only $D5$ has good chances of being universally true in the sense that it holds for all processes. In sum, dialectics is not a universal doctrine: it is not true of all things, all properties, and all changes. It can be exemplified but also counterexemplified. (See also Miró Quesada, 1972.)

It follows, in particular, that dialectics fails to embrace all objects, whether physical or conceptual. More exactly, I submit that dialectics, to the extent to which it can be regarded as an ontology of physical objects, does not apply to conceptual objects and therefore is not a generalization of formal logic. Moreover there can be no universal theory holding for both physical objects and conceptual objects: constructs (concepts, propositions, theories) satisfy conceptual laws which, unlike the laws of nature, are manmade.

We do not find constructs ready-made in nature, nor do we manufacture them out of physical things: constructs are figments of our brain's creative activity and they are characterized by laws of their own, which do not apply to physical objects. Thus a proposition does not move and does not get wet or rusty, just as a piece of iron cannot become negated and cannot entail another physical object. Propositions are characterized by the propositional calculus, sets by set theory, groups by group theory, and so on, whereas concrete objects are characterized by physical (or biological or sociological) laws. There is little in common between the two sets of laws, the physical and the conceptual. To be sure one may speak of the conjunction (e.g. juxtaposition) of two bodies as well as of the conjunction of two propositions. But physical conjunction is not defined the same way as logical conjunction. In particular, the De Morgan laws make no sense for physical conjunction, if only because there is no such thing as the negation of a physical object (except when interpreted as its environment).

In short, it is not possible to cover all objects, whether physical or conceptual, by a single theory. In particular dialectics cannot do this job. Since dialectics can be exemplified by some physical entities and events, but is at variance with formal logic, it should be regarded as an ontological (or metaphysical or cosmological) theory. Even so its scope is, as we have seen, rather narrow.

4.9. THE RELATION OF DIALECTICS TO FORMAL LOGIC

Dialecticians have claimed that logic is a special case of dialectics, namely a sort of slow motion approximation that holds when change is exceedingly slow hence negligible to a first approximation. This claim is false. Indeed, for one law statement to be a particular case of another, both must refer to the same things. And this is not the case with the laws of logic and the hypotheses of ontology. On the other hand the laws of classical electrodynamics are a limiting case (for large number of photons) of the laws of quantum electrodynamics: both sets of laws are comparable because both concern radiation. This is not the case with the predicate calculus, or any other logical theory, in relation to the laws of physics or of ontology: whereas the former dscribes the behavior of concepts and propositions, the latter is concerned with describing physical systems. And, as we have seen in Section 4.8, constructs (unlike the processes of thinking of them) are not physical objects. Formal logic, then, cannot be a particular case of dialectical ontology. What then is the relation between the two?

The relation between logic and any cogent nonlogical (e.g. ontological) theory is not a relation of reduction but the relation of presupposition. Logic is presupposed (logically, not psychologically or historically) by all other cogent theories. So much so that, when performing the orderly reconstruction (i.e. axiomatization) of any substantive theory, whether in mathematics or in science or in ontology, one must start by specifying the language in which the theory is to be couched as well as the rules of inference the theory will admit or use. In short, the logic must be specified in advance. Moreover, the logic is not altered if the theory is found inconsistent with experience, because the former does not concern experience but rather our mode of organizing experience (including purely mental experience).

Logic proper, i.e. the set of logical theories, is subject matter free and immune to experience: it is an a priori frame serving just as well for mathematics as for physics or for sociology. Since any substantive theory T presupposes some logical theory L, T contains or entails L. T is said to be a *substantive* theory because it describes some nonlogical objects, such as numbers or people, while L is indifferent to the precise reference. Indeed the concepts and propositions occurring in L may refer to anything, hence to nothing in particular. On the other hand if T is impoverished to the point that all of its substantive assumptions are removed, what remains is a skeleton without any precise reference: at most, T may then be said to refer to individuals in some unspecified abstract set. But this is not precisely what dialectics is supposed to be about.

To put it in another way: formal logic *refers* to everything but describes or *represents* nothing but its own basic concepts — "not", "and", "for all", "entails", and their kin. These specific concepts of logic refer or apply to propositions, not to material objects. In fact, consider e.g. the logical connective "or", which may be construed as a function mapping pairs of propositions into propositions. (In symbols, $\vee: P \times P \to P$, where P is the set of propositions.) And consider on the other hand the ontological concept

of interaction, which cannot be regarded as applying to propositions. Indeed, "interacts" relates concrete objects: more precisely,
the interaction predicate is one that takes pairs of concrete objects
into propositions such as "*a* interacts with *b*". (In symbols, I:
$C \times C \rightarrow P$, where C is the set of concrete objects.) If we admit that
the reference class of a predicate equals the union of all the sets
occurring in its domain of definition (Bunge, 1974a), we obtain

$$\mathscr{R}(\vee) = P, \qquad \mathscr{R}(I) = C.$$

And, since propositions are disjoint from concrete or physical objects, the two predicates have nothing in common except their
general form, which is a mathematical property (namely the property of being binary predicates). In other words, logic on the one
hand and physics (whether *stricto sensu* or *lato sensu*, i.e. ontology) on the other, do not concern the same objects. Hence neither
can be a special case of the other.

The previous argument assumes of course that physical objects
are disjoint from conceptual objects, in particular propositions
(i.e. $P \cap C = \emptyset$). This hypothesis cannot be proved but it may be
rendered plausible, namely thus. Whereas things are out there, in
the external word, constructs have no physical or material existence: they exist only conceptually, i.e. as members of conceptual
bodies (e.g. theories). When stating that *there are* constructs of
some kind, e.g. that the number 3 exists, or that *there is* Schrödinger's equation, all we intend to convey is this: We think out
certain ideas and pretend that they thereby acquire an independent existence, i.e. that they have become independent of their
psychological genesis and historical development. We take this
kind of existence seriously: otherwise we would be unable to do
logic and mathematics and we would be unable to distinguish them
from psychology and history. But, unless we are Platonists or
Hegelians, we do not assign ideas a separate or autonomous existence. Only living thinkers have a concrete or physical existence

and so do their brain processes. We just feign or pretend that what they think (their ideas) can be detached (imaginarily, not physically) from their thought processes. So much so that we endow ideas with nonphysical properties such as having a sense and a truth value. (More in Chapter 9.)

The thesis of the oneness of logic and ontology is possible, nay necessary, in an idealist system, where there can be no radical difference between things and constructs except that the latter are deemed to be superior to the former. Hegel's conflation of logic with ontology was thus quite natural — in his system. It is also natural to a vulgar materialist or nominalist, because he does not admit concepts but only their symbols. But to anyone who is neither an idealist nor a vulgar materialist logic will be distinct from ontology. This does not entail an ontological dualism as long as constructs are not assigned an independent existence. But it does entail a *methodological dualism* according to which constructs are treated *as if* they had an autonomous existence — which of course they don't really. (It should come as no surprise to find that fictionalism is true of fictions though not of reality.)

The idea that the understanding of change requires a logic of its own, be it dialectical logic or some version of temporal logic, because formal logic is incapable of dealing with change, is a relic from ancient philosophy. It was justifiable two millenia ago, when people could not ask more precise questions than "Is the arrow moving or is it at rest?", and were puzzled by the question whether the arrow in flight was or was not at a given place at a given instant. Nowadays we think in degrees rather than in opposites, asking instead "How fast is the car moving relatively to the ground?", and are not baffled by the possible answer "The car moves with zero velocity", which Parmenides might have regarded as self-contradictory. Moreover we do not regard these questions as logical but as scientific, and have become accustomed to handling mutually incompatible substantive theories with the help of one and the same logic. In short *we no longer think dialectically*, i.e. in opposites and without distinguishing logic from the disciplines dealing with facts. So when we fail to understand some kind of change we blame our failure on some substantive theories, not on logic, which is one of the tools employed in building, testing and criticizing scientific theories: logic is a priori.

4.10. BALANCE SHEET

The upshot of our examination of dialectical ontology is this:

(i) The principles of dialectics, as formulated in the extant literature, are ambiguous and imprecise. It behoves the student of dialectics to elucidate the key notions involved and to reformulate those principles in a clear and cogent fashion.

(ii) When formulated more carefully, all but one or two of the principles of dialectics lose their universality: they start with the prefix "Some" rather than "All". And, when stated in this weaker form, some of them become so narrow as to border on platitudes — for example, the hypothesis that there are systems with mutually opposed components.

(iii) Even when formulated clearly and with a restricted scope, the principles of dialectics do not constitute a sufficient base for a theory of change. They constitute at most an embryo that might be developed into a theory proper. Genuine modern theories of change should be far more precise, explicit and complete than that. Besides, they ought to be congenial with science rather than at variance with it. In particular, they should not be couched in archaic terms such as 'the struggle of opposites' — except of course when authentic conflicts among genuine opposites happen to be at stake.

(iv) Dialectics does not embrace formal logic, if only because the latter is concerned with constructs not with the real world. The claim that dialectics does generalize logic can be upheld only within a Platonic or a Hegelian ontology and is incompatible with any realist epistemology, in particular with naive realism or the reflection theory of knowledge. (By the way, we do not accept the latter.)

(v) The two correct principles of dialectics — that every real thing is in a state of flux, and that new properties are bound to emerge (or to get lost) along any process — are shared by all process ontologies. Moreover they can be formulated in an exact manner as well as linked with other general ontological hypotheses to form a consistent hypothetico-deductive system in harmony with science. (See Bunge, 1977a, 1979.) This new ontology is dynamicist but not dialectical. If preferred it keeps, elucidates and systematizes what is alive, but discards what is dead, in dialectics.

(vi) Because it has been formulated in vague and metaphorical terms, dialectics is hard to confront with facts, i.e. difficult to test for truth. Moreover dialecticians do not take kindly to counter-examples and tend to brand criticisms of dialectics as 'undialectical' or 'metaphysical'. But this gambit is a tacit avowal that the "laws" of dialectics are not universal after all. For, if they were universal, if they were laws of nature and society and thought, then nothing undialectical, whether in the brain or without, should ever happen: we should all think and behave dialectically under every circumstance, and thus the teaching of dialectics should be unnecessary.

(vii) Because dialectics promises so much and delivers so little, it has fostered superficiality, particularly in the social sciences. Indeed many a writer in this field feels that he has accomplished something by declaring that a certain process is dialectical, or that it is one more confirming instance of dialectics. Compare such instant wisdom with the procedure of a scientist who, upon suspecting that he has to do with opposing things or properties of things, proposes precise hypotheses concerning the mode and mechanism of such opposition, checks whether his hypotheses jibe with the background knowledge, translates such conjectures into the language of mathematics, draws some deductive consequences, and collects or produces data to check those hypotheses. (Elementary example of a hypothesis concerning the rates of growth or decline, $\dot{A}$ and $\dot{B}$, of the properties A and B respectively of a single thing or of two components of a system: $\dot{A} = a_{11}AB - a_{12}B$, $\dot{B} = a_{21}AB - a_{22}A$, where the coefficients are positive real numbers to be determined empirically. If $a_{11} = a_{21} = 0$, pure competition; if $a_{12} = a_{22} = 0$, pure cooperation; if neither of the coefficients is nought, a mixture of competition and cooperation – a far more common occurrence than dialecticians would like.)

(viii) Because it is a component of systems of thought that purport to account for everything, dialectics is usually embraced with partisan enthusiasm and zeal, and criticisms of it rejected out of hand. As a consequence dialectics has stagnated: it has become a dogma. True, there have been a few feeble attempts to free dialectics from dogmatism, but they have retained the same 'spirit': imprecision, superficiality, unwarranted claims of universality, and disregard for counterexamples. A system of dogmas cannot be much improved by partial reformations or revisions: it calls for a fundamental revolution touching upon principles as well as upon the very way principles have to be proposed, systematized, tested, and discussed. In this regard dialecticians have proved to be conservative, not revolutionary.

(ix) Dialectics is intellectually damaging because (a) it excuses inconsistency, (b) it consecrates archaic modes of thinking, particularly thinking in opposites rather than in degrees, and (c) it alienates its believers from science, which acknowledges a great many modes of becoming, as well as from the method of science, which involves criticism and empirical test.

(x) Dialectics is not just one more philosophical dogma but one with dangerous practical consequences, for it renders people obsessed with conflict and prone to engage in it, blinding them to the possibility and benefit of cooperation.

PART THREE

MIND

CHAPTER 5

A MATERIALIST THEORY OF MIND

Materialism answers, among others, the question of the nature of mind, i.e. the mind-body problem. However, so far this answer has been sketchy and inexact, and it has not made full use of contemporary physiological psychology. The aim of this chapter is to provide a fuller and more precise answer on the basis of some of the recent work in neuroscience and psychobiology.

5.1. TEN VIEWS ON THE MIND-BODY PROBLEM

There are two main sets of solutions to the problem of the nature of mind: psychoneural monism and psychoneural dualism. While according to the former mind and brain are one in some sense, according to dualism they are separate entities. However, there are considerable differences among the components of each of the two sets of solutions to the mind-body problem. Thus psychoneural monism is composed of the following alternative doctrines: panpsychism ("Everything is mental"), neutral monism ("The physical and the mental are so many aspects or manifestations of a single entity"), eliminative materialism ("Nothing is mental"), reductive materialism ("The mind is physical"), and emergentist materialism ("The mind is a set of emergent brain functions or activities"). Likewise the dualist camp is divided into five sects: autonomism ("Body and mind are mutually independent"), parallelism ("Body and mind are parallel or synchronous to one another"), epiphenomenalism ("The body affects or causes the mind"), animism ("The mind affects, causes, animates or controls the body"), and interactionism ("Body and mind interact").

None of these views is too clear: none of them is a theory proper, i.e. a hypothetical-deductive system with clearly stated assumptions, definitions, and logical consequences from them. Every one of the above opinions on the nature of mind has been formulated only in verbal terms and with more concern for obeisance to ideology than for the data and models produced by neuroscientists and psychologists. In particular, although there are plenty of arguments pro and con the so-called *identity theory*, or materialist theory of mind, nobody seems to have produced such a comprehensive theory in the strict sense of the term 'theory'. All we have, in addition to a number of psychophysiological models of a few special mental functions, is a programmatic hypothesis — namely that mind is a set of brain functions. To be sure, this hypothesis has had tremendous heuristic power in guiding research in the neurophysiology of mental processes. Yet it is insufficient, for scientists need a more explicit formulation of the thesis that what "minds" is the brain, and philosophers would find it easier to evaluate the claims of the psychoneural identity "theory" if it were stated with some precision and in some detail.

The present chapter attempts to accomplish just that with regard to one kind of psychoneural identity theory, namely emergentist materialism. This is the view that mental states and processes, though brain activities, are not just physical or chemical or even cellular, but are specific activities of complex neuron assemblies. These systems, evolved by some higher vertebrates, are fixed (Hebb, 1949) or itinerant (Craik, 1966; Bindra, 1976). This chapter is based on another, more comprehensive and formal work (Bunge, 1980), which in turn uses key concepts elucidated elsewhere (Bunge, 1977a, 1979) — particularly those of system, biosystem, and biofunction. Only the bare bones of the theory are presented here.

5.2. BASIC CONCEPTS AND HYPOTHESES

A basic concept of our theory is that of a concrete or material system, as exemplified by a neuronal circuit, a neuronal mini-column, and the entire CNS of an animal. A concrete system can be characterized by its composition, environment, and structure. (Recall Chapter 2, Section 2.3.) We are of course particularly interested in nervous systems and their subsystems, so we start by proposing

DEFINITION 1. A system is a *nervous system* iff it is an informa-tion biosystem such that
   (i) it is composed of (living) cells;
   (ii) it is or has been a proper part of a multicellular animal;
   (iii) its structure includes (a) the regulation or control of some of the biofunctions of the animal, and (b) the detection of internal and environmental events as well as the transmission of signals triggered by such events.

DEFINITION 2. A biosystem is a *neural* (or *neuronal*) system iff it is a subsystem of a nervous system.

DEFINITION 3. A biosystem is a *neuron* iff it is a cellular com-ponent of a neural system.

DEFINITION 4. Let v be a neural system and $\mathscr{C}_t(v)$ the neuronal composition of v at time t. Further, call

$$C_t: \ \mathscr{C}_t(v) \times \ \mathscr{C}_t(v) \rightarrow [-1, 1]$$

the real valued function such that $C_t(a, b)$, for $a, b \in \mathscr{C}_t(v)$, is the strength (intensity) of the connection (coupling, link) from neuron a to neuron b at time t. Then the *connectivity* of v at t is represented by the matrix formed by all the connection values, i.e.,

$$\mathbb{C}_t = \|C_t(a, b)\|.$$

DEFINITION 5. A connectivity is *constant* iff it does not change once established (i.e. iff $\mathbb{C}_t$ is independent of time). Otherwise it is *variable*.

DEFINITION 6. A neuronal system is *plastic* (or *uncommitted*, or *modifiable*, or *self-organizable*) iff its connectivity is variable throughout the animal's life. Otherwise (i.e. if it is constant from birth or from a certain stage in the development of the animal), the system is *committed* (or *wired-in*, or *prewired*, or *preprogrammed*).

DEFINITION 7. Any plastic neural system is called a *psychon*.

Our initial assumptions are as follows.

POSTULATE 1. All animals with a nervous system have neuronal systems that are committed, and some animals have also neuronal systems that are plastic (uncommitted, self-organizable).

POSTULATE 2. The neuronal systems that regulate (control) the internal milieu, as well as the biofunctions of the newborn animal, are committed (wired-in).

POSTULATE 3. The plastic (uncomitted) neuronal systems of an animal (i.e. its psychons) are coupled to form a supersystem, namely the *plastic neural supersystem* (P) of the animal.

POSTULATE 4. Every animal endowed with psychons (plastic neuronal systems) is capable of acquiring new biofunctions in the course of its life.

DEFINITION 8. Every neural function involving a psychon (or plastic neuronal system) with a regular connectivity (i.e. one that is constant or else varies regularly) is said to be *learned*.

Notice that this is a neurophysiological not a behavioral definition of learning, and moreover one in line with the use-disuse hypothesis (Hebb, 1949), which is becoming increasingly popular among neuroscientists.

We proceed now to refine the above ideas with the help of the concepts of state function and state space, well known to physicists and general system theorists, and which are quickly invading all of the sciences.

Consider an arbitrary neuronal system, be it a small neuronal circuit (fixed or itinerant), a sizable subsystem of the CNS, or the entire brain of a higher vertebrate. Like every other system, it can be represented by a state function

$$\mathbb{F} = \langle F_1, F_2, \ldots, F_n, \ldots \rangle,$$

every component of which is assumed to represent a property of the system. If we let the connectivity function (Definition 4) take care of the spatial distribution, we can assume that $\mathbb{F}$ is only a time-dependent function whose values may be taken to be n-tuples of real numbers. I.e. we can set $\mathbb{F}: T \to \mathbb{R}^n$, with $T \subseteq \mathbb{R}$, and $\mathbb{R}$ equal to the set of reals.

Each component F_i of the total state function $\mathbb{F}$ of the neuronal system concerned can be decomposed into a constant (or nearly constant) part $\mathbb{F}_i^c$ and a variable part $\mathbb{F}_i^v$. Obviously, either can be zero during the period in question. However, the point is that, while the rate of change of $\mathbb{F}^c$ vanishes at all times (i.e. $\dot{\mathbb{F}}^c = 0$), that of $\mathbb{F}^v$ does not. The latter may thus be taken to represent the activity of the neuronal system, i.e. what it does. For this reason we make

DEFINITION 9. Let $\mathbb{F}: T \to \mathbb{R}^n$ be a state function for a neuronal system ν, and let $\mathbb{F} = \mathbb{F}^c + \mathbb{F}^\nu$, with $\dot{\mathbb{F}}^c = 0$ (zero rate of change) for all t in the interval T. Then

(i) ν is *active* at time t iff $\mathbb{F}^\nu(t) \neq 0$;

(ii) the *intensity* of the activity of ν over the time lapse $\tau \subset T$ equals the fraction of the components of ν active during τ;

(iii) the *state of activity* of ν at time t is $s = \mathbb{F}^\nu(t)$;

(iv) the (total) process (or *function*) ν is engaged in over time interval $\tau \subset T$ is the set of states of activity of g:

$$\pi(\nu, \tau) = \{\mathbb{F}^\nu(t) \,|\, t \in \tau\}.$$

DEFINITION 10. Let $\pi(\nu, \tau)$ be the total process (or function) of a neuronal system ν in an animal b during the time interval $\tau \subset T$. The corresponding *specific function* (activity, process) of ν during τ is whatever ν but none of the other subsystems of b does during τ, i.e.,

$$\pi_s(\nu, \tau) = \pi(\nu, \tau) - \bigcup_{\mu \prec b} \pi(\mu, \tau), \quad \text{with } \mu \neq \nu.$$

(If A and B are sets then $A - B = A \cap \overline{B}$, where $\overline{B}$ is the complement of B. The union is taken over every subsystem of b, i.e. for all $\mu \prec b$.)

We shall now introduce the hypothesis that the CNS, and every neuronal subsystem of it, is constantly active even in the absence of external stimuli:

POSTULATE 5. For any neural system ν of an animal, the instantaneous state of activity of ν decomposes additively into two functions: $\mathbb{F}^\nu = \mathbb{A} + \mathbb{E}$, where $\mathbb{A}$ does not vanish for all $t \in T$, whereas $\mathbb{E}$ depends upon the actions of other subsystems of the animal upon ν.

DEFINITION 11. Let $\mathbb{F}^\nu = \mathbb{A} + \mathbb{E}$ be the active part of the state function of a neural system v. Then $/A(t)$ is the *state of spontaneous activity* of v at time t, and $\mathbb{E}(t)$ the *state of induced* (or *stimulated*) *activity* of v at t.

5.3. MENTAL STATES AND PROCESSES

Every fact experienced introspectively as mental is assumed to be identical with some brain activity: this, in a nutshell, is the biological or materialist view of mind. For example, vision is the activity of neural systems in the visual system, learning the formation of new neural connections, intending the activity of certain neuronal systems in the forebrain, and so on. But not all brain activity is mental: we assume that the mental is the specific function of certain plastic neuronal systems (all of which discharge also household functions such as protein synthesis). Our assumption takes the form of

DEFINITION 12. Let b an animal endowed with a plastic neural system P. Then

(i) b undergoes a *mental process* (or performs a mental function) during the time interval τ iff P has a subsystem v such that v is engaged in a specific process during τ;

(ii) every state (or stage) in a mental process of b is a *mental state* of b.

Example. Acts of will are presumably specific activities of neuron modules in the forebrain. *Nonexample*. Hunger, thirst, fear, rage, and sexual urge are processes in subcortical systems (mainly hypothalamic and limbic), hence are nonmental. What is a mental process is the consciousness of any such states — which is a process in some subsystem of P.

COROLLARY 1. All and only animals endowed with plastic

neural systems are capable of being in mental states (or undergoing mental processes).

COROLLARY 2. All mental disorders (dysfunctions) are neural disorders (dysfunctions).

COROLLARY 3. Mental functions (processes) cease with the death of the corresponding neural systems.

COROLLARY 4. Mental functions (processes) cannot be directly transferred (i.e. without any physical channels) from one brain to another.

DEFINITION 13. Let P be the plastic (uncommitted) supersystem of an animal b of species K. Then
  (i) the *mind* of b during the period τ is the union of all mental processes (functions) that components of P engage in during τ:

$$m(b, \tau) = \bigcup_{x < P} \pi_s(x, \tau);$$

  (ii) the *K-mind*, or *mind of species K*, during period τ, is the union of the minds of its members during τ:

$$M(K, \tau) = \bigcup_{y \in K} m(y, \tau).$$

THEOREM 1. *The mental functions of (processes in) the plastic neural supersystem of an animal are coupled to one another, i.e. they form a functional system. (The unity of the mind principle.)*
  Proof: By Postulate 3 the components of P, far from being uncoupled, form a system.

COROLLARY 5. Let b be an animal whose plastic neural system is split into two detached parts, L and R. Then the mind of b

during any time period posterior to the splitting divides into two disjoint functional systems:

$$m(b, \tau) = m_L(b, \tau) \cup m_R(b, \tau), \quad \text{with}$$
$$m_L(b, \tau) \cap m_R(b, \tau) = \emptyset.$$

THEOREM 2. *Mental events can cause nonmental events in the same body and conversely.*

Proof: Mental events are neural events, and the causal relation is defined for pairs of events in concrete things.

So much for generalities. Let us move on to specifics.

5.4. SENSATION AND PERCEPTION

DEFINITION 14. A detector is a *neurosensor* (or *neuroreceptor*) iff it is a neural system or is directly coupled to a neural system.

DEFINITION 15. A *sensory system* of an animal is a subsystem of the nervous system of it, composed of neurosensors and of neural systems coupled to these.

DEFINITION 16. A *sensation* (or *sensory process*) is a specific state of activity (or function or process) of a sensory system.

DEFINITION 17.
(i) A *percept* (or *perceptual process*) is a specific function (activity, process) of a sensory system and of the plastic neural system(s) directly coupled to it;
(ii) a *perceptual system* is a neural system that can undergo perceptual processes.

We assume that the perception of an external object is the distortion it causes on the ongoing activity of a perceptual system:

POSTULATE 6. Let ν be a perceptual system of an animal b, and call $\pi_s(\nu, \tau) = \{ \mathbb{F}^\nu(t) \mid t \in \tau \}$ the specific process (or function) that ν engages in during the period τ when in the presence of a thing x external to ν, and call $\pi_s^0(\nu, \tau)$ the specific function of ν during τ when x fails to act on ν. Then b perceives x as the symmetric (Boolean) difference between the two processes. I.e. the perception of x by b during τ is the process

$$p(x, \tau) = \pi_s(\nu, \tau) \Delta \pi_s^0(\nu, \tau),$$

where $A \Delta B = (A - B) \cup (B - A)$, i.e. everything in A but not in B plus everything in B but not in A.

We perceive events, i.e. changes of state. And not just any events but those originating in some neurosensor or acting on the latter and, in any case, belonging to our own event space (or the set of changes occurring in us). And our perceptions are in turn events in the plastic part of our own sensory cortex. Normally these are not fully autonomous events but events that map or represent events occurring in other parts of the body or in our environment. To be sure, this mapping is anything but simple and faithful, yet it is a mapping in the mathematical sense, i.e. a function. Thus we assume

POSTULATE 7. Let b be an animal equipped with a perceptual system c, and call $S(b)$ the state space of b, and $S(c)$ that of c. Moreover let $E(b) \subset S(b) \times S(b)$ be the animal's event space, and $E(c) \subset S(c) \times S(c)$ its perceptual event space. Then there is a set of injections (one to one and into functions) from the set of bodily events in b to the set of perceptual events in c. Each such map, called a *body schema*, depends on the kind of bodily events as well as on the state of the animal. I.e. the general form of each body map is

$$m: S(b) \times 2^{E(b)} \longrightarrow 2^{E(c)},$$

where 2^X is the family of all subsets of the set X.

POSTULATE 8. Let $E(e)$ be a set of events in the environment e of an animal b equipped with a perceptual system c, and call $S(b)$ the state space of b and $S(c)$ that of c. Moreover let $E(b) \subset S(b) \times S(b)$ be the animal's event space, and $E(c) \subset S(c) \times S(c)$ its perceptual event space. Then there is a set of partial maps k from sets of external events in $E(e)$ to ordered pairs ⟨state of b, set of bodily events in b⟩, and another set of partial maps p, from the latter set to sets of perceptual events. Furthermore the two sets of maps are equally numerous, and each map k composes with one map p to form an *external world map* of b in e, or ϵ. I.e.

$$\epsilon : 2^{E(e)} \xrightarrow{\ k\ } S(b) \times 2^{E(b)} \xrightarrow{\ p\ } 2^{E(c)} .$$

DEFINITION 18. Let b be an animal with perceptual system c in environment e. Moreover call $S(b)$ the state space of b and $E(e)$ the set of events in e. Then b, when in state $s \in S(b)$, *perceives* external events in $x \in 2^{E(e)}$ if, and only if, [these cause bodily events that are in turn projected on to the sensory cortex c, i.e. if] $k(x) = \langle s, y \rangle$ with $y \in 2^{E(b)}$ and in turn $p(s, y) \in 2^{E(c)}$. Otherwise the events in x are *imperceptible* to b when in state s [i.e. imperceptible events either do not cause any bodily events or cause them but do not get projected on to the perceptual system].

5.5. BEHAVIOR

We turn now from perception to motor outputs, i.e. behavior, starting with

DEFINITION 18. For any animal b,

(i) the *behavioral state* of b at time t is the state of motion of b at t;

(ii) the *behavior* of b during the time interval τ is the set of all behavioral states of b throughout τ.

DEFINITION 19. A *behavior pattern* is a recurrent behavior.

DEFINITION 20. Let b be an animal of species K, and let A be the union of all animal species. Then

(i) the (possible) *behavior of type i of animal b*, or $B_i(b)$, is the set of all (possible) behaviors of b associated with the ith biofunction (in particular neural biofunction) of b;

(ii) the *behavioral repertoire* of animal b, or $B(b)$, is the union of all (possible) behavior types of b, i.e.

$$B(b) = \bigcup_{i=1}^{n} B_i(b);$$

(iii) the (*possible*) *behavior of type i of species K*, or $B_i(K)$, is the union of all the (possible) behaviors of the members of K, i.e.

$$B_i(K) = \bigcup_{x \in K} B_i(x);$$

(iv) the *behavioral repertoire of species K*, or $B(K)$, is the union of all (possible) behavior types of K:

$$B(K) = \bigcup_{i=1}^{n} B_i(K);$$

(v) the *specific behavioral repertoire of species K* is the behavioral repertoire exclusive to members of K:

$$B_s(K) = B(K) - \bigcup_{X \subset A} B(X), \quad \text{with} \quad X \neq K;$$

(vi) *animal behavior* is the union of the behavioral repertoires of all animal species, i.e. $B = \bigcup_{X \subset A} B(X)$.

We assume that behavior, far from being primary, is derivative:

POSTULATE 9. The behavior of every animal endowed with a nervous system is produced ("mediated", "subserved") by the latter. I.e. for every behavior type B_i of animals endowed with a nervous system, the latter contains a neural subsystem producing the motions in B_i.

COROLLARY 6. Any change in (nonredundant) neural systems is followed by some behavioral changes.

THEOREM 3. *No two animals behave in exactly the same manner.*
   Proof: By Postulate 9 and the general ontological principle that there are no two exactly identical things.

THEOREM 4. *The behavioral repertoire of an animal endowed with plastic neural systems splits into two parts: the one controlled by the committed (or prewired) part of the NS of the animal, and its complement, i.e. the behavior controlled by the plastic components of the NS.*
   Proof: By Postulates 1 and 4 together with Definition 20.

DEFINITION 21. The part of the behavioral repertoire of an animal that is controlled by the committed part of its NS, is called its inherited (or *instinctive, stereotyped, modal,* or *rigid*) repertoire, while the one controlled by the plastic part of its NS, its *learned* repertoire.

COROLLARY 7. The behavior of an animal deprived of plastic neural systems is totally stereotyped.

POSTULATE 10. Provided the environment does not change radically during the lifetime of an animal, most of its inherited behavioral repertoire has a positive biovalue for it.

POSTULATE 11. Some of the inherited capabilities of an animal endowed with plastic neural systems are modifiable by learning.

So much for our general behavioral principles. Now for motivation.

5.6. MOTIVATION

DEFINITION 22. A *drive* (or *motivation*) of kind X is the detection of an imbalance in the X components(s) of the state function of the animal. (More precisely: the intensity $D_X(b, t)$ of drive X in animal b at time t equals the absolute value of the difference between the detected and the normal values of X for b at t.)

POSTULATE 12. For every drive in an animal there is a type of behavior of the animal that reduces that drive (i.e. that decreases the imbalance in the corresponding property and thus tends to bring the animal back to its normal state).

We come finally to values:

DEFINITION 23. Let S be a set of items and b an animal. Further, let $\succsim_b$ be a partial order on S. Then the structure $V_b = \langle S, \succsim_b \rangle$ is a *value system* for b at a given time iff
  (i) b can detect any member of S and discriminate it from all other items in S;
  (ii) for any two members x and y of S, b either prefers x to y $(x \succsim_b y)$ or conversely $(y \succsim_b x)$ or both $(x \sim_b y)$ at the given time.

POSTULATE 13. All animals are equipped with a value system, and those capable of learning can modify their value systems.

DEFINITION 24. Let $V_b = \langle S, \succsim_b \rangle$ be a value system for an animal b at a given time, and call $A \subset S$ a set of alternatives open to b, i.e. belonging to the behavioral repertoire of b at the time. Then b *chooses* (or *selects*) option $x \in A$ iff

(i) it is possible for b to pick (i.e. to do) any alternative in A (i.e. b is free to choose);

(ii) b prefers x to any other options in A; and

(iii) b actually picks (i.e. does) x.

Note the difference between preference and choice, obscured by the operationalist doctrine: preference underlies and motivates choice, which is valuation in action.

5.7. MEMORY AND PURPOSE

Many systems besides animals have memory, so the following definition is quite general:

DEFINITION 25. A system σ at time t has *memory* of (or *memorizes*) some of its past states iff the state of σ at t is a function(al) of those past states.

POSTULATE 14. All animals have memory of some of their past states, and none of all of them.

DEFINITION 26. Call P a kind of event or process in a neural system of an animal b involving a plastic subsystem, and S a kind of stimuli (external or internal) which b can detect. Then b has *learned* $p \in P$ in the presence of $s \in S$ during the time interval $[t_1, t_2]$ iff

(i) p did not occur in b in the presence of s before t_1;

(ii) after t_2, p occurs in b whenever b senses s [i.e. b has memorized p].

Since all behavior is controlled by some neural system (Postulate 9), the previous definition embraces the concept of behavioral learning, i.e. acquisition of new behavior patterns in response to new environmental situations.

DEFINITION 27. The *experience* of an animal at a given time is the set of all it has learned up until that time.

So far we have dealt with non-anticipatory systems. We now introduce anticipation, an ability only few species possess:

DEFINITION 28. Animal *b* *expects* (or *foresees*) a future event of kind *E* when sensing an (external or internal) stimulus *s* while in state *t*, iff *b* has learned to pair *s* and *t* with an event of kind *E*.

Animals capable of anticipatory behavior can act purposively:

DEFINITION 29. An action *X* of an animal *b* has the *purpose* or *goal Y* iff
   (i) *b* may choose not to do *X*;
   (ii) *b* has learned that doing *X* brings about, or increases the chance of attaining, *Y*;
   (iii) *b* expects the possible occurrence of *Y* upon doing *X*;
   (iv) *b* values *Y*.

The conditions of purposiveness are then freedom, learning, expectation, and valuation. Since machines do not fulfil all four conditions, they cannot be goal-seeking except by proxy.

DEFINITION 30. An action *X* of an animal *b* is a *suitable means* for attaining a goal *Y* of *b* iff in fact *b*'s performing *X* brings about, or increases the probability of occurrence of, *Y*.

5.8. THINKING

Let us now tackle concept attainment and proposition formation.

We assume that forming a concept of the "concrete" kind — i.e. a class of real things or events — consists in responding uniformly to any and only members of the given class:

POSTULATE 15. Let C be a set of (simultaneous or successive) things or events. There are animals equipped with psychons whose activity is caused or triggered, directly or indirectly, by any member of C, and is independent of what particular member activates them.

DEFINITION 31. Let C be a class of things or events, and b an animal satisfying Postulate 15, i.e. possessing a psychon that can be activated uniformly by any and only a member of C. Then b *attains a concept* $\theta_b(C)$ of C (or *conceives* C, or *thinks up* C) iff the activity (process, function) stimulated by a C in that psychon of b equals $\theta_b(C)$.

We now conjecture that forming a proposition is the chaining of the psychons (possibly cortical columns) thinking up the concepts occurring in the proposition:

POSTULATE 16. Thinking up a proposition is (identical with) the sequential activation of the psychons whose activities are the concepts occurring in the proposition in the given order.

POSTULATE 17. A sequence of thoughts about propositions is (identical with) the sequential activation of the psychons whose activities are the propositions in the sequence.

We can now characterize the various modes of knowing:

DEFINITION 32. Let a be an animal. Then
(i) if b is a learned behavior pattern, a *knows how to do* (or perform) b iff b is in the actual behavioral repertoire of a;

(ii) if c is a construct (concept, proposition, or set of either), then a *knows* c iff a thinks up (or conceives) c;

(iii) if e is an event, then a has *knowledge* of e iff a feels or perceives e, or thinks of e.

5.9. DECISION AND CREATIVITY

We can use the concept of knowledge to elucidate that of decision:

DEFINITION 33. Let x be an arbitrary member of a set A of alternatives accessible to an animal b with the value system $V_b = \langle S, \gtrsim_b \rangle$. Then b *decides to choose* x iff

(i) b has knowledge of some members of A;

(ii) $A \subset S$ (i.e. b prefers some members of A to others);

(iii) b in fact chooses x.

The ability to make decisions is then restricted to animals capable of knowing. And rational decision is of course even more restricted:

DEFINITION 34. A decision made by an animal is *rational* iff it is preceded by

(i) adequate knowledge and correct valuations, and

(ii) foresight of the possible outcomes of the corresponding action.

DEFINITION 35. A *rational animal* is one capable of making some rational decisions.

Finally, a few words about creativity, a characteristic of all higher vertebrates.

DEFINITION 36. Let a be an animal of species K with behavioral repertoire $B(K)$ at time t. Then

(i) a *invents* behavior type (or pattern) b at time $t' > t$ if a does b for the first time, and b did not belong to $B(K)$ until t';

(ii) a *invents* construct c at time $t' > t$ iff a knows c for the first time at time t';

(iii) a *discovers* event e at time $t' > t$ iff a has knowledge of e for the first time at time t';

(iv) a is *creative* iff a invents a behavior type or a construct, or discovers an event before any other member of its species;

(v) a is *absolutely creative* (or *original*) iff a creates something before any other animal of any species.

POSTULATE 18. Every creative act is the activity, or an effect of the activity, of newly formed psychons. [New connections, not new neurons.]

POSTULATE 19. All animals endowed with plastic neural systems are creative.

5.10. SELF TO SOCIALITY

We start by drawing a sharp distinction between awareness and consciousness:

DEFINITION 37. If b is an animal,

(i) b is *aware* of (or notices) stimulus x (internal or external) iff b feels or perceives x — otherwise b is *unaware* of x;

(ii) b is *conscious* of brain process x in b iff b thinks of x — otherwise b is unconscious of x.

POSTULATE 20. All animals are aware of some stimuli and some are also conscious of some of their own brain processes.

DEFINITION 38. The *consciousness* of an animal b is the set of all the states of the CNS of b in which b is conscious of some CNS process or other in b.

POSTULATE 21. Let P be a subsystem of the CNS of an animal b engaged in a mental process p. Then the CNS of b contains a neural system Q, other than P and connected with P, whose activity q equals b's being conscious (thinking) of p.

DEFINITION 39. An animal
(i) has (or is in a state of) *self-awareness* iff it is aware of itself (i.e. of events occurring in itself) as different from all other entities;
(ii) has (or is in a state of) *self-consciousness* iff it is conscious of some of its own past conscious states;
(iii) has a *self* at a given time iff it is self-aware or self-conscious at that time.

POSTULATE 22. In the course of the life of an animal capable of learning, learned behavior, if initially conscious, becomes gradually unconscious.

DEFINITION 40. An animal act is *voluntary* (or *intentional*) iff it is a conscious purposeful act — otherwise it is *involuntary*.

DEFINITION 41. An animal acts of its own *free will* iff
(i) its action is voluntary and
(ii) it has free choice of its goal(s) — i.e. is under no programmed or external compulsion to attain the chosen goal.

THEOREM 5. All animals capable of being in conscious states are able to perform free voluntary acts.
Proof: By Postulate 20 and Definitions 40 and 41.

DEFINITION 42. If *b* is an animal endowed with a plastic neural system capable of mentation (i.e. with a nonvoid mind), then

(i) the *personality* of *b* is the functional system composed of all the motor and mental functions of *b*;

(ii) a *person* is an animal endowed with a personality.

Finally we face sociality:

DEFINITION 43. An animal engages in *social behavior* iff it acts on, or is acted upon by, other individuals of the same genus.

POSTULATE 23. The behavioral repertoire of every animal includes types (patterns) of social behavior.

5.11. CONCLUDING REMARKS

The greatest merit of the psychoneural identity theory is that it takes it for granted that mind can be investigated scientifically, while the greatest sin of psychoneural dualism is to deny this or at least to make things difficult for the physiological psychologist who investigates the brain in order to understand mentation and behavior.

If we wish to approach the study of mind in a scientific manner we must start by identifying the concrete system in question, be it a neuron, neuron assembly, or entire nervous system. Now, physiological psychology, psychoendocrinology and psychopharmacology tell us that the thing in question, that which does the minding and controls the behaving, is the CNS, in particular the brain. In this approach the mind is not an entity apart from the brain, parallel to it or interacting with it. In the psychobiological approach mind is a collection of activities of the brain or some multimillion neuronal subsystem of it. For example the felt intensity of a stimulus (the subjective experience we have of it) is conjectured to be identical with the frequency of firing of certain neural systems (including the corresponding cortical area).

All this can be stated clearly with the help of the state space formalism, which is not just a formal trick but a method used, or utilizable, in all factual sciences because it fits in with the ontology of lawfully changing concrete things. The dualist cannot adopt this method because there is no way of merging properties of the brain with properties of an immaterial substance to form a single vector spanning a single state space. So, if the dualist attempted to speak a mathematical language — which he does not — he would be forced to split the state space of a person into two different and separate spaces, one of which would be ill-defined for it would be characterized in verbal and nonphysiological terms.

The rejection of psychoneural dualism does not force one to adopt eliminative materialism (or physicalism). Psychobiology suggests not only psychoneural monism — the strict identity of mental events with brain events — but also emergentism, or the thesis that mentality is an emergent property possessed only by animals endowed with a very complex plastic nervous system. This ability confers its owners such decisive advantages, and is related to so many other properties and laws (physiological, psychological, and social), that one is justified in asserting that the organisms endowed with it constitute a level of their own — that of psychosystems. But of course this is not saying that minds constitute a level or "world" of their own, and this simply because there are no disembodied (or even embodied) minds, but only minding bodies.

In other words, one can hold that the mental is emergent relative to the merely physical or chemical without reifying the former. That is, one can hold that the mind is not an entity composed of lower level things — let alone an entity composed of no entities whatever — but a collection of functions (activities, processes) of certain neural systems, that individual neurons do not possess. And so emergentist (or systemist) materialism, unlike eliminative materialism, is seen to be compatible with overall pluralism, or the ontology that proclaims the qualitative variety and mutability of reality (Bunge, 1977c).

Our espousing emergentist (or systemist) materialism and proposing a general framework for handling the mind-body problem does not entail claiming that the latter has been solved. It hasn't and it won't, except in very general terms, for emergentist materialism is a philosophy providing only a scaffolding for the detailed scientific (empirical and theoretical) investigation of the many particular problems lumped under the rubric 'the mind-body problem'. It behoves neuroscientists and psychologists to attack these problems — as scientists, not as amateur philosophers, let alone theologians. They won't do so if told that mind is a mysterious immaterial entity best left in the hands of philosophers and theologians.

CHAPTER 6

MIND EVOLVING

If materialism is false, the human mind need not develop from infancy nor evolve from prehuman stages: being immaterial, mind need not accompany the vicissitudes of matter. But if emergentist and evolutionary materialism is true, then mind must develop and evolve along with brain: it must grow from infancy as the central nervous system matures, and it must acquire new abilities as primates evolve through *Homo erectus, H. habilis, H. sapiens*, and finally into *H. sapiens sapiens*. The study of the development and evolution of mind is therefore a good testing ground for materialism and its rivals. This chapter is devoted to the second problem, to wit, the evolution of brain functions.

6.1. ONE CENTURY OF EVOLUTIONARY PSYCHOLOGY

Until recently the mind was generally regarded as a human prerogative as well as immaterial, unchanging, and often also as supernatural. Charles Darwin changed all this, if not *de facto* at least *de jure*. In fact he conjectured that subhuman animals too can have a mental life, that ideation is a bodily process, and that it is subject to natural selection just like any other biofunction.

Darwin adopted a materialist and evolutionist view of mind as early as 1838, as revealed by his recently published M and N Notebooks, which he edited himself the same year (1856) as he began writing *Natural Selection*, the unfinished ancestor of the *Origin of Species* (Gruber and Barrett, 1974). In the M Notebook we read: "Origin of man now proved. – He who understands baboon would do more toward metaphysics than Locke" (M84). And in

91

the N Notebook he draws the methodological consequence of the
hypothesis that mind is a bodily function: "To study Metaphysics,
as they have always been studied appears to me to be like puzzling
at astronomy without mechanics. — Experience shows the problem
of the mind cannot be solved by attacking the citadel itself — the
mind is function of body — we must bring some *stable* foundation
to argue from —" (N5).

Darwin of course did more than confide his thoughts on the
nature and evolution of behaviour and ideation to his notebooks.
He wrote *The Descent of Man* (1871) and *The Expression of the
Emotions in Animals and Men* (1872), the foundation stones of
evolutionary psychology and ethology — not to speak of palaeo-
anthropology and prehistory. The first book, in particular, exerted
a tremendous influence — on all but psychologists and philosophers
of mind. True, George Romanes wrote several books on animal
intelligence and mental development (which he misleadingly called
'evolution'). However, those works fell short of scientific rigour.
Besides, Romanes rejected mental evolution on religious grounds.
Likewise Alfred Russel Wallace, the cofounder of evolutionary
biology, resisted Darwin's generalization of evolution to the
human brain and its mental abilities: to him these could have only
a supernatural explanation. (Compare the current resistance of
some psychologists to the idea that apes can be taught certain
languages).

The field lay fallow until C. Lloyd Morgan published his *Introduction to Comparative Psychology* (1894). From then on psychologists have taken it for granted that the study of pigeons, rats,
dogs and monkeys is relevant to the understanding of the human
species. But unlike the behaviourists who followed (in particular
Watson and Skinner), Morgan believed not only in the continuity
of the evolutionary process but also in the emergence of new properties and laws during it: he wrote *The Emergence of Novelty*
(1933) to emphasize that point. And the distinguished animal

psychologist Theodore Schneirla emphasized the qualitative differences among evolutionary stages in his long essay "Levels in the
psychological capacities of animals" (Schneirla, 1949).

Like most other evolutionists, the eminent geneticist Theodosius
Dobzhansky thought that evolution had been punctuated by qualitative novelties. In particular he stated that man is "a possessor of
mental abilities which occur in other animals only in most rudi-
mentary forms, if at all" (Dobzhansky, 1955). This acknowledge-
ment of qualitative difference drew the irate reaction of a distinguished primatologist, who stated that "There is no evidence
of an intellectual gulf [between man and subhuman animals] at
any point" (Harlow, 1958). And yet, despite all the astonishing
accomplishments of apes – particularly in artificial environments
– none of them has designed a machine, or written a song, or
proposed a social reform. So, abyss there is, though it was formed
only as recently as about 5 million years and in one and the same
mountain range.

6.2. THE PERSISTENT NEGLECT OF
EVOLUTIONARY PSYCHOLOGY

Evolutionary psychologists will then stress the discontinuities
(through emergence) as well as the continuity (through descent)
of the evolutionary process. Hence, although they will regard rat
psychology as highly relevant to human psychology, they will
also investigate the peculiarities of human behaviour and ideation.
But how many behavioural scientists actually think in evolutionary terms? A perusal of scientific publications will exhibit but few
works on the evolution of the nervous system aside from the
monumental volume on the evolution of the cerebellum edited
by Llinás in 1969 (Llinás, 1969). Only a handful of books on the
evolution of behaviour and mentation appeared after the volume
edited by Roe and Simpson in 1958: Munn (1971), Jerison (1973),

and Masterton and colleagues (1976). And although there are hundreds of journals of psychology, with a few papers on evolutionary psychology scattered among them, none seems exclusively devoted to evolutionary psychology — a term that is not even in current usage.

It seems then that most psychologists are not used to thinking in evolutionary terms or, indeed, in biological terms. For the most part they have received no biological training and they seldom mix with biologists. (Neuroscientists retaliate by ignoring psychology.) As a result, although there are thousands of departments of biology and departments of psychology in the world, there are only a handful of departments of psychobiology (or neuropsychology, or neuroethology, or biobehavioral science). Moreover it is only recently that courses in physiological psychology have become standard components of psychology curricula. Even so, few textbooks in the field adopt an explicit evolutionary viewpoint (an exception is Thompson 1975).

The disregard for evolutionary biology is particularly notable among the practitioners of artificial intelligence (AI). Intent as they are on designing machines imitating human behaviour or ideation, they often take simulation for identity and claim that human beings are machines. By thus skipping the biological and social levels they gloss over three billion years of evolution. And by so doing they fall into the dualist trap: indeed the AI people are fond of likening the mind-body distinction to the software-hardware distinction. They also forget that, unlike self-assembled (or self-organizing) systems, all machines have been designed and none of them is free in any sense, for every one of them acts by proxy. A pinch of evolutionary thought would have spared them these mistakes — and deprived the dualist of the joy of receiving comfort from the mechanist.

If psychologists on the whole still ignore evolution, it is little wonder that philosophers of mind pay hardly any attention to it.

A well-known Wittgensteinian philosopher of mind noted for his attacks on the psychoneural identity hypothesis has gone so far as to hold that "it is really senseless to conjecture that animals *may* have thoughts" (Malcolm, 1973). Yet we have known for some years now, particularly thanks to the work of the Gardners, Premacks, and Rumbaughs, that apes can learn to express their thoughts by means of certain artificial languages such as that of the deaf-mutes.

There is no justification for a philosopher to continue to talk of mind as a human prerogative as well as immaterial and unchanging, let alone as supernatural. There is even less justification for a neuroscientist to continue to hold such prescientific views on mind. However this is exactly what the eminent neurophysiologist Sir John Eccles has been saying for the past three decades: that mind is immaterial yet can act on neurons (Eccles, 1951, 1980); and that the existence of consciousness and that of the cosmos "require a supernatural explanation to be admitted by we scientists in all humility" (Eccles, 1978). Not surprisingly, Eccles claims that *Homo sapiens* has not been subject to evolution. In fact, since the brain of man (seems to have) ceased to grow over the past 250,000 years, "our human inheritance of brains averaging about 1400 cc in size is the end of the evolutionary story. And now, in any case, biological evolution for man is at an end because selection pressure has been eliminated by the welfare state" (Eccles, 1977).

This startling conclusion rests on the following false presuppositions: (a) that genic variation (by mutation and recombination) stopped long ago in human beings; (b) that behaviour and ideation, though possibly dependent upon brain size, do not depend upon the organization and plasticity of the neurons; (c) that behaviour and ideation play no active role in adaptation, hence in human evolution; (d) that hominids and humans have never faced new challenges (such as glaciations and droughts) calling for new abilities; (e) that the Welfare State, by alleviating poverty, raising the health level and assuring education for all, does not set a premium on any skills.

In sum Eccles has managed to do violence to genetics, evolutionary biology, history and sociology in one breath. (For further criticism see Dimond, 1977.) This is not a coincidence but an unavoidable consequence of the dualistic and supernaturalistic philosophy of mind according to which either mental evolution does not exist or, if it does, then it is not an aspect of biological (and social) evolution. Evolutionary neuroscientists think otherwise: to them "The relation between brain and mind evolves in association with a physiological substratum, and hence there is no dualism" (Bullock, 1958). See details in Masterton, 1976.

6.3. RESEARCH PROBLEMS SPARKED OFF BY PSYCHONEURAL MONISM

It would seem that we are back to square one, in the pre-Darwin days, when biologists and philosophers of mind held a palaeolithic view of mind. But this is not quite true, for psychoneural monism has now a growing number of adepts, and evolutionary thinking is making important inroads in psychology, particularly among primatologists, developmental psychologists and sociobiologists. These scientists are at least beginning to pose problems that make no sense in a dualistic and a pre-evolutionary context, and that are spawning some interesting hypotheses and experiments. Here is a random sample:

(1) How new is the neocortex and, in particular, the associative "area"?

(2) How new is synaptic plasticity — or, equivalently, when did Hebb's mechanism of use and disuse emerge? (Hebb's principle is "Neurons that fire together tend to stay together").

(3) What is the origin of brain lateralization? Is there any adaptive advantage in it? (See, for example, Levy, 1977.)

(4) Why does parental behaviour occur at all, given that it is energy- and time-consuming and that it renders parents easy targets for predators (Barash, 1976)?

(5) What has been the evolution of parental behaviour in the higher vertebrates and how is it affected by ideation?

(6) Is altruism genetically programmed or learnt? (Prior question: Is it legitimate to treat inheritance and experience as additive factors and consequently ask what the contribution of each is?)

(7) Is aggression genetically programmed? (Prior questions: the one in point (6), and whether aggression can be equated with defence and making a living.)

(8) When and how did empathy and solidarity emerge?

(9) What are the origins of goodness and of cruelty?

(10) How did smiling and laughter evolve from grimacing, and what if any are their adaptive advantages?

(11) At what evolutionary level did imagination, ideation, foresight (or foresmell), goal-seeking behaviour and self-consciousness start?

(12) How did human language originate and evolve? (See Jaynes, 1976; Parker and Gibson, 1979.)

(13) Is acquisitiveness instinctive or learnt, and how did it evolve?

(14) Is there an inherited moral sense, and in any case at what point in evolution did morality emerge? (Recall Darwin's *Notebook M* (144): "What is the Philosophy of Shame & Blushing? Does Elephant know shame − dog knows triumph." − For recent work on the development of morality see Goslan, 1969.)

(15) What if any are the biological roots of managerial, political and military behaviour patterns?

The above are just a few of the many questions that are currently being investigated, or are likely to be tackled soon, by evolutionary psychologists, ethologists, sociobiologists, anthropologists and social scientists. No such questions arise in the framework of psychophysical dualism even if dualists occasionally pay lip service to evolutionary biology (e.g. Popper and Eccles, 1977). The reason is transparent: according to the neo-Darwinian theory, evolution proceeds by genic variation (a material event) and natural selection (another material event). That theory leaves no room for immaterial agencies such as non-embodied souls, ideas in themselves, and other Platonic inmates of Popper's World 3. (See Chapter 8.)

Dualists do not accept materialistic (biological) explanations in the matter of mind: they cannot admit that molecular changes and environmental factors can change minds. So, they must either deny mental evolution altogether or speculate that it proceeds by some mechanism other than genic variation and natural (and social) selection. In either case they come into conflict with evolutionary biology and psychology, regardless of the frequency with which they use the term 'evolution'. And, to the extent that their views are influential, dualists slow down the advance of both sciences.

Psychoneural identity theorists, on the other hand, are in harmony with evolutionary biology and psychology. In fact to them mental functions are brain functions, so mental evolution is an aspect of the evolution of animals possessing brains capable of "minding" — i.e. the higher vertebrates. (See Chapter 5 and, for details, Bunge, 1980.) To be sure, for social animals, in particular human beings, we must add social (economic, political and cultural) evolution to biological evolution. But, since societies are concrete or material (though not just physical) systems, biosocial evolution is a material process — albeit one exhibiting properties unknown to physics or even biology. This is not to deny the influence of behaviour and ideation on adaptation, and hence on evolution. On the contrary, the efficacy of behaviour and ideation is assured by construing them as bodily functions, whereas if detached from matter they are deprived of testable power.

In sum, psychoneural monism is the philosophy of mind behind evolutionary (and physiological) psychology. And, while the latter has yet to make its mark, at least it has made a good start by asking some of the proper questions — questions whose investigation does not call for the postulating of immaterial or supernatural agencies.

4. OBSTACLES: GENUINE AND SPURIOUS

Several objections have been raised against the neo-Darwinian approach to behaviour and ideation. One of them is that, because behaviour is not only a result of evolution but also a factor in it, the theory must be corrected with a dose of Lamarckism (Piaget, 1976). This claim seems to be mistaken: Behavioural (and mental) adaptation can be explained by (natural or artificial) selection acting on genic variation. The environment selects behaving organisms, not naked genes or disembodied behaviour repertoires.

To be sure, the higher vertebrates can learn new behaviour patterns in response to environmental changes or mental processes, so they are not at the mercy of their genomes — or, rather, they can make full use of their genetic potentialities. Still, such potentialities are inherited and, when actualized, they work on a par with other biological traits. If successful, a new behaviour pattern — whether acquired by genic change or by learning — is bound to have some genetic effect because it favours certain genotypes over others.

Example 1. A mutant synthesizes certain enzymes, allowing it to eat certain plants that the normal variety cannot digest. This allows the mutant to occupy a different (perhaps larger) ecological niche, which confers some advantage upon it though at the same time setting it in competition with other species. If successful, the mutation will tend to spread.

Example 2. An animal learns a new advantageous behaviour pattern, i.e. one that eases its way of making a living. As a consequence it reproduces earlier or more abundantly than its conspecifics, so that its peculiar genes have better chances of spreading and becoming entrenched. In either case the rate of change of the new population depends upon the difference between the birth and death rates — as is the case with all organisms — but now these rates depend partly on behavioural traits, some of which are phenotypic rather than genotypic.

So, while behaviour can make the difference between radiation and extinction, it fits within the neo-Darwinian schema. What is true is not that the latter is grossly in error but that behaviour, in particular social behaviour, enriches the modes of adaptation and quickens considerably the pace of evolution. (Methodological consequence: genetics is necessary but insufficient to explain behavioural evolution.) This applies particularly to the plastic behaviour of the higher vertebrates, which is partly controlled by ideation. The overall advantage (and occasional disadvantage) of possessing mental abilities is quite obvious in an evolutionary psychobiological perspective. If on the other hand mind were immaterial, evolutionary theory would have nothing to say about it.

Another popular idea is that human history transcends biological history (true) to the point that the latter is irrelevant to the former (false). This is like saying that, because biology deals with emergent properties and laws ignored by physics, the latter is irrelevant to the former. Surely human history is more than biological evolution: it encompasses the latter, in the sense that human history is artificial rather than natural evolution, for history is largely shaped (though not fully controlled) by human beings themselves. But this history is concrete and thus a far cry from the mythical history of independently evolving ideas imagined by Hegel and other idealist philosophers. Human history is concrete because it is the evolution of human populations, i.e. of systems of human beings interacting amongst themselves and with their natural and social environments.

What is undoubtedly true is that the weight of intelligence in human affairs has increased dramatically over the past fifty millennia and even more so since the beginning of agriculture about 10,000 years ago, and far more so since the emergence of literacy about 5,000 years ago. However, intelligence is a property of the brain – and one extremely sensitive to social stimuli – not of an immaterial mind. And in any case let us not forget that, with the increasingly important role of intelligence in the process of artificial selection that we call 'human history', the roles of human stupidity, greed and cruelty have also become more important. Side by side with the inventor, the scientist and the social reformer, human societies have created the dogmatist, the mystic and the military butcher. While more and more individuals have become enlightened during certain exceptional periods, greater and greater masses have been driven to disaster, and now a couple of individuals have the technical ability to extinguish all life on earth – several times over. There is no room for such monstrous stupidity in the idealistic philosophy of history.

A third popular objection to evolutionary psychology is that it is untestable, for a collection of fossil bones can tell us nothing about the behaviour and mentation of their departed owners. This objection is quite natural if behaviour and mentation are taken to be immaterial, but it is no more than a challenge if the psychobiological standpoint is adopted. Moreover the challenge is being met by the usual methods elaborated by palaeontologists, anthropologists and prehistorians. There are three such methods and, while all three involve admirable efforts of imagination, neither surpasses the imagination invested by physicists in conjecturing the structure of the atom or by biochemists in guessing the structure of the DNA molecule.

There is, first, the direct approach of studying fossils and their associated natural and artificial remains, and trying to reconstruct not only their anatomy but also the physiology and the behaviour of their owners (in particular posture, locomotion and feeding habits). *Example:* The hypothetical reconstructions of the ways of life of hominids living two million years ago in East Africa. (For a beautiful semipopular account see Leaky and Lewin, 1977.)

Then there is the comparative approach, which is both indirect and empirical, as it consists in studying specimens of modern taxa conjectured to be close relatives of extinct ones. (Not all comparative studies are relevant to evolutionary psychology even when they focus on behaviour. Only those are relevant which compare the behavioural repertoires and mental abilities of species belonging to the same phyletic line. See Hodos and Campbell, 1969.) *Example:* Primatologists have been shedding considerable light on the possible life styles and behavioural repertoires of our prehuman ancestors.

Finally there is the armchair approach of asking how, on the basis of our general neurobiological knowledge, certain neuronal systems and their functions could have evolved. All three approaches are necessary and should be better integrated, particularly in the case of behaviour and ideation, for they throw light on complementary sides and control each other. (Cf. Jerison, 1973.)

In sum, although evolutionary psychology is a tough subject, it may not be tougher than cosmology and, in any case, it is just as interesting if not more so. The same holds for evolutionary epistemology (Campbell, 1974; Vollmer, 1975).

6.5. SUMMARY AND CONCLUSIONS

Our first conclusion is that evolutionary psychology, created by Darwin one century ago, is a going concern, admittedly still rather weak. It deserves the strong support of all scientists and philosophers interested in extending the scientific approach to the study of mind.

Secondly, the underdeveloped state of evolutionary psychology may be explained by (a) the youth of evolutionary thinking in general, (b) the neglect of the theory of evolution (and in general of biology) by most psychologists, (c) the difficulty of generating reasonable hypotheses about the evolution of behavior and ideation — particularly given the strong prejudice against framing hypotheses and theories in the behavioral sciences, (d) the difficulty of obtaining and interpreting empirical data relevant to such hypotheses, and (e) last, but not least, the dualistic philosophy of mind that has dominated for thousands of years.

Thirdly, evolutionary biology has done more than trace the ancestry of some contemporary biospecies: it has reoriented biological thought. Thus where pre-Darwinians saw design in adaptation, contemporary biologists see the outcome of natural or artificial selection on the products of random genic variation. The impact of the new outlook on the sciences of man has been just as dramatic. Thus while the pre-Darwinians often saw man as a spiritual being (Plato's model), we see ourselves as animals, albeit exceptional ones (Aristotle's model). While the pre-Darwinians (in particular the believers in the arcane) believed man to have paranormal abilities independent of the brain (such as telepathy and psychokinesis), we believe all our mental functions to be lawful (though not always normal) brain functions. And while the pre-Darwinians (in particular the psychoanalysts) sought a purpose or "meaning" in every bit of human behavior and mentation, we regard purposeful behavior as exceptional and moreover as something to be explained rather than as having an explanatory force. *Example:* "She did X in order to attain Y" is explained as either "She has been conditioned to do X every time she wanted Y", or "Knowing (or suspecting) that X causes Y, and valuing Y, she did X".

Fourthly, when scientists underrate philosophical issues they risk falling victim to unscientific philosophies likely to slow down or even derail the train of their research – as Engels (1872–82) noted long ago. The mind-body problem is a case in point: as an eminent psychologist wrote three decades ago, "The study of mental evolution has been handicapped by a metaphysical [mind-body] dualism", for it denies the evolutionary hypothesis that "The evolution of mind is the evolution of nervous mechanisms" (Lashley, 1949). Indeed (a) instead of suggesting promising questions that may be investigated with the means at the disposal of biologists, anthropologists and prehistorians, dualism diverts the attention of these investigators to an inscrutable entity – the proverbial immaterial soul; (b) because it contends that mind is immaterial, dualism must either deny that it has evolved or assert that its evolutionary mechanism is non-Darwinian: in either case it is anti-evolutionary.

Fifthly, in contrast to psychoneural dualism, the psychoneural identity theory suggests a host of interesting research problems, such as those of the forms "What neural systems are required by such and such behavior patterns or mental functions?", and "How may such and such behavior patterns or mental capacities have evolved?" Yet, the psychoneural identity theory is so far a skeleton waiting to be fleshed out, and evolutionary psychology is nothing but an incipient science that can boast of far more problems than solutions. But at least these problems are interesting and can be investigated by the scientific method, and those few solutions that we do have are not ready-made ideological slogans but scientific hypotheses that can be perfected or replaced.

Finally, if epistemology is taken to be the study of cognitive processes, not just of their products (knowledge), then it cannot help being both genetic (i.e. developmental) and evolutionary. But the very concepts of development (or ontogenesis) and evolution (or phylogenesis) of mind make no sense in a dualist context, let alone in an idealist one. Only a materialist philosophy of mind, and particularly emergentist materialism, offers genetic and evolutionary epistemology a congenial philosophical background. Indeed, according to emergentist materialism the appearance and refinement of cognitive abilities, be it in the individual or in the species, far from being mysterious, is an aspect of the development or the evolution of the brain interacting with the rest of the body as well as with its natural and social environment.

PART FOUR

CULTURE

CHAPTER 7

A MATERIALIST CONCEPT OF CULTURE

The very existence of culture has long been regarded as a challenge
to materialism or even a refutation of it. How could a sonata, a
poem, a theorem or a theory be regarded as a material entity?
Most materialists have responded to the challenge by granting
more or less tacitly that a culture is substantially different from
material objects and have added that, notwithstanding, culture
is determined by material factors such as the environment and the
economy. However, the very admission that culture is not material
is suicidal for the materialist. And what is the point of being a
materialist with regard to matter but a dualist with regard to culture? I shall not adopt the economy-culture duality and shall construct instead a materialist concept of culture consistent with the
full-blooded materialist ontology sketched in Chapter 2. But
before doing so we shall have to remove a terminological hurdle.

7.1. SOCIETY AND CULTURE

Contemporary anthropologists often use the term 'culture' where
sociologists, economists, historians and sociobiologists employ the
word 'society'. Thus anthropologists will speak of the Maya culture, or of the Brooklyn culture, instead of the Maya society or of
the Brooklyn society. And they may say, as they told me, that
digging a latrine is just as cultural as writing a poem.

This peculiar use of the word 'culture' is unfortunate for the
following reasons. Firstly, it is idiosyncratic and therefore creates
an unnecessary barrier between anthropology and the other
sciences. Secondly, if adhered to consistently it forces us to em-

ploy quaint expressions, such as 'the culture of a culture', when clearer phrases, like 'the culture of a society' are available. Thirdly, it prevents us from employing useful expressions such as 'the economy of a culture' (meaning the economic component of cultural activities). Fourthly, the substitution of 'culture' for 'society' suggests that culture, in the strict sense, overrides everything else, i.e. kinship, economy and polity. Fifthly, the said equivocation is an obstacle to the very formulation of the question whether certain nonhuman societies – e.g. baboon and chimpanzee societies – have a culture, or at least the rudiments of it – i.e. a language, a tradition, and the capacity for discovery and invention.

Any one of the above five reasons should suffice for dropping the identification of culture with society, and adopting instead a strict sense of 'culture', i.e. that of a subsystem of society. Thus the phrase 'the Maya culture' will be short for 'the culture of the Maya society' – a companion to 'the kinship system of the Maya society', 'the economy of the Maya society' and 'the polity of the Maya society'. And the Maya culture will be understood to embrace Maya art and architecture, poetry and drama, astronomy and time keeping, arithmetic and botany, and so on – but not Maya agriculture and trade nor Maya political organization and international relations. The Maya culture is then a part of the Maya society, not the whole of it. But what part of a society is its culture? I.e. how can the latter be characterized? Let us turn to this question.

Since a culture is characterized by certain activities, we may as well draw up a short haphazard list of typical cultural activities in human societies:

> communication by spoken or written word
> training of the young
> game playing, dancing and singing
> drawing, painting and sculpting

composing and performing music
story-telling and writing, and play-acting
witchcraft, magic, and religion
land surveying
time keeping
nature lore
healing
making conjectures and putting them to the test
inventing tools and manufacturing processes
doing mathematics and science
speculating and arguing
chronicling.

On the other hand gathering fruit, roots or eggs, hunting, building, manufacturing tools, trading, overseeing work or having accepted mores observed, administering, warring, and the like are noncultural activities. However, all such activities are guided (or misguided) by beliefs and values belonging to the culture: even the mere gathering of food is done in the light of both positive knowledge and superstition concerning plants and their virtues. In other words, although we can and must *distinguish* between cultural and noncultural activities we must not *separate* them.

All cultural activities are of course activities of individuals, whether acting singly or in cooperation with others. Hence there are no such things as creative writing or mathematical research by themselves: there are only creative writers and original mathematicians. Literature apart from writers or mathematics apart from mathematicians are just useful fictions – nay, indispensable ones when it comes to analyzing the "products" of such activities. But those are not the concern of anthropologists, concerned as they are with real people and not with abstractions. In sum, a culture is composed not of fields such as literature and mathematics but of *people* doing literature, mathematics, and so on.

The anthropology of culture, then, just like that of economy or that of polity, is concerned with people engaged in cultural activities. Such activities are performed by individuals but not by isolated ones: even solitary contemplation is done by individuals who are knitted in a social fabric and who have been trained or at least influenced by others. In other words, *cultural activities are social even when they are conducted by single individuals.* The same holds, a fortiori, for economic and political activities. Therefore the anthropologist is concerned with individuals and organizations as they influence, or are influenced by, other individuals or organizations.

Cultural activities are of course not the only social activities. The activities and connections that hold a society together — or that eventuate in its breakdown — can be split into four classes: biological, economic, political, and cultural. Thus courting, mating and rearing of young are biological activities. Trade is an economic relation or, equivalently, trading is an economic activity. On the other hand state control of trade is a political activity — in the wide sense of 'political' adopted here. And designing a sewage network, or a biological experiment, are cultural activities — not free, of course, from economic and political constraints.

No social activity is purely biological, purely economic, purely political, or purely cultural, except as far as its goals is concerned. For example, trading involves the use of language (if only of sign language), which is a cultural item. Likewise proving a theorem may involve the use of pencil and paper, which are products of economic activity, which is in turn subject to political controls. In general, every sector of social activity — be it the economy, the culture of the polity — involves people and artifacts from the other two sectors. In particular, cultural activities or outputs are the result of cultural, political, and economic inputs.

Social activity is then highly *systemic*: whatever happens in one sector is likely to affect events in the other sectors. This is of course well known to the functionalists, the materialists, and those who approach society from a general systems point of view. However, the idea that the study of social activities calls for a systems approach is not popular enough or it is often mistaken for the holistic abhorrence of analysis. Let us therefore sketch a systems framework for the study of social activities and, in particular, cultural activities.

7.2. SOCIETIES AS SYSTEMS AND CULTURES AS SUBSYSTEMS

A concrete system is an aggregate of concrete components that share an environment and are bonded together. The minimal characterization of a concrete system consists then in listing its composition, its environment, and the bonds that hold the components together. Since all three components or coordinates of such a triple are likely to change in the course of time, the above list must be indexed by time − i.e. we must speak of the composition, environment, and bonds of a system at a given time. In what follows this will be taken for granted.

If society is to be conceived of as a concrete system then the coordinates of the triple must be specified. We submit that every society σ, at any given moment of its existence, can be represented schematically by listing the following items:

(i) the *composition* or *membership M* of σ;

(ii) the *immediate environment* (natural or social) of σ, i.e. the set E of items that, not being in M, act on members of σ or are acted upon by them;

(iii) the *structure* of σ, i.e. the set S of social (interpersonal and intergroup) relations or activities of members of σ, plus the set T of relations and activities of transformation of environmental items, involving members of M.

This schema can be symbolized as the ordered triple

$$s = \langle M, E, S \cup T \rangle,$$

where '$\cup$' designates the set theoretic union. The first coordinate of s is of course what the social scientist is ultimately interested in, namely people — not however isolated people but individuals interacting with one another and with the environment. Any study of σ should begin then by identifying its membership M rather than dealing with disembodied behavior patterns or attitudes or beliefs or values or what have you. From this point of view, 'Society σ lives in an arid zone' is short for 'The members of σ live in an arid environment'. Likewise 'Society σ practises pottery' abbreviates 'Some members of σ make pottery', where making pottery is one of the relations in the set T of transformation relations. And 'Society σ values education' must be understood as 'Most members of σ seek to educate or to be educated', where educating is one of the relations in the set S of social relations.

Every one of the three coordinates of s must be nonempty if the society is to be other than ghostly. And none of the three coordinates exists by itself. A society in an environmental vacuum is just as imaginary as a society devoid of social structure, i.e. one which neither transforms its environment nor is held together by social relations. Therefore any reasonable approach to the study of a society requires taking all three coordinates into account. This seems self-evident, yet idealism ignores real people, environmentalism neglects the social structure, societalism forgets about the environment, and structuralism dispenses with all three. (In particular, according to Lévy-Strauss social structure is not objective but a figment of the anthropologist's imagination.)

Needless to say, building a schema of society only begins the study of it. A second step in the study of a society consists in distinguishing its subsystems, a third in building models of the latter, and a fourth in tracing the evolution of the system and its subsystems. Let us deal with the matter of the subsystems. But first a definition: A thing is a *subsystem* of a system (called the *main system*, or *supersystem*) if and only if it is a part of the system and is a system itself. For example, a family is a subsystem of modern society. On the other hand the members of a family, though parts of the latter, are not social systems and therefore do not qualify as subsystems of society.

The very first thing to do in trying to discover the subsystems of a social system is to find out what its members actually do besides keeping alive. Now, whatever the type of society, its active population can be split into three main subgroups:

(i) the *labor force* ^1P, or the people engaged mainly in agriculture, industry, trade, or service activities;

(ii) the *cultural force* ^2P, or the people engaged mainly in cultural activities;

(iii) the *managerial force* ^3P, or the people engaged in controlling whatever economic or cultural activities are carried out in the society.

This classing of the membership of a society by occupational groups is natural in a way, but rather artificial from the systems-theoretic viewpoint because neither of the three manpowers listed above is a system, i.e. a concrete thing behaving as a unit in some respect by virtue of its inner structure. In fact, the members of every one of the above groups are scattered among different sectors or spheres. Thus while some of the intellectuals, artists, teachers, priests, etc. belonging to the cultural force of a given society are attached to cultural organizations (e.g. schools), others work in economic organizations (as, e.g., government physicians or geologists or writers). The more advanced a society, the more intense such a diffusion of the specialized personnel among the various sectors of the society.

We are thus led to shifting the focus from occupational groups (the manpowers ^1P, ^2P, ^3P) to the subsystems they compose. These subsystems, we submit, are three for every society, no matter how primitive or developed: the economy, the culture, and the polity.

7.3. CHARACTERIZATION OF THE THREE MAIN ARTIFICIAL SUBSYSTEMS OF SOCIETY

We submit then that the ith manpower ^iP of each society is distributed among the three main subsystems and thus is divided into the following: a part ^iP_E engaged in some facet of economic production, another part ^iP_C engaged in some facet of cultural production, and a third part ^iP_G engaged in political (e.g. governmental) organization. In short,

$$^iP = {}^iP_E \cup {}^iP_C \cup {}^iP_G, \quad \text{for} \quad i = 1, 2, 3.$$

Whereas the upper indices name the type of occupation (e.g. worker, intellectual, manager), the lower indices denote the primary goal of the activity. In other words, the upper indices name the inputs and the lower ones the primary or specific output of the subsystem concerned. Thus the production of a book may call for the cooperation of printers and binders (members of ^1P_E), author and editors (members of ^2P_C), as well as governmental cultural agents such as grants officers and censors (members of ^3P_G).

Correspondingly we distinguish certain subsets of the collection S of social relations in any given society: we call S_E, S_C, and S_G the social relations of material production, cultural production, and political administration respectively. Finally we perform a similar splitting of the work relations or activities: we call L_E, L_C and L_G the kinds of work done in economic production, in the cultural sphere and in government respectively by members of the labor force; C_E, C_C and C_G, the kinds of cultural work involved in economic production, in the cultural sector and in politics respectively; and M_E, M_C, and M_G the kinds of management involved in economic production, cultural production, and politics respectively. We are now ready for

DEFINITION 1. Let $s = \langle M, E, S \cup T \rangle$ represent a society σ with labor force ^1P, cultural force ^2P, and managerial force ^3P. Further, let the subindices E, C, and G identify whatever is associated with agricultural or industrial production or services, cultural production, and political administration respectively. Finally, call Q_i the subset of the membership M of σ related to the manpower ^iP, e.g. consumers, recipients of cultural production, or victims of political oppression. Then

(i) the subsystem of σ represented by

$$\epsilon_\sigma = \langle {}^1P_E \cup {}^2P_E \cup {}^3P_E \cup Q_E, E, S_E \cup (L_E \cup C_E \cup M_E) \rangle$$

is called the *economic system* (or *economy*) of σ;

(ii) the subsystem of σ represented by

$$\kappa_\sigma = \langle {}^1P_C \cup {}^2P_C \cup {}^3P_C \cup Q_C, E, S_C \cup (L_C \cup C_C \cup M_C) \rangle$$

is called the *cultural system* (or *culture*) of σ;

(iii) the subsystem of σ represented by

$$\pi_\sigma = \langle {}^1P_G \cup {}^2P_G \cup {}^3P_G \cup Q_G, E, S_G \cup (L_G \cup C_G \cup M_G) \rangle$$

is called the *political system* (or *polity*) of σ.

This definition allows us to formulate two trivial yet necessary assumptions: (i) *Every society is composed of three main artificial subsystems: the economy, the culture, and the polity;* (ii) *every sociosystem (i.e. organization) in any society is part of at least one of the main subsystems of the society.*

The above conception of society elicits the following comments. Firstly, by Theorem 1 of Chapter 2, Section 2.3, *every social system, in particular every culture, is a material entity* because it is composed of material (though not just physical) objects, namely animals of some species. Secondly, by conceiving of society as a material system – and only in this way – one makes sense of phrases such as 'the energy flow through society' and 'the interaction between society and environment'. Thirdly, by conceiving of the economy, the culture, and the polity as systems, one avoids the sterile philosophies of holism and of individualism. By the same token one obtains the possibility of modeling the entire economy, the entire culture and the entire polity of a society as systems with definite compositions and structures. Furthermore one can recognize input variables and output variables as well as inner mechanisms. (More on this in Section 7.5.) Fourthly, all three artificial subsystems share the same immediate (natural or artificial environment; in particular, there is no cultural system functioning in a void. Hence extreme internalism (e.g. cultural idealism) is seen to be just as inadequate as extreme externalism (e.g. ecological determinism). Fifthly, each manpower is distributed among the three subsystems. In particular, not even the most primitive economic system fails to employ some cultural workers and organizers – even if they are at the same time primary producers – and not even the most anti-intellectual of political régimes dispenses altogether with intellectuals – if only to control the creative thinkers and artists. Sixthly, it is possible – nay desirable – for one and the same individual to belong to two or more of the main subsystems, as is the case when the division of labor is embryonic. Seventhly, a significant change in any one of the three coordinates of each system will affect the other. (This is of course what the functionalists call 'interdependence among social functions'.) Examples: a sudden shortage, or expansion, of energy sources; a sudden increment, or decrement, in the number of skills; a rapid increase, or decrease, in the number and quality of intellectuals. Any such drastic change in one component is bound to affect the other two. (And when the overall change is as large as it is fast, it is called a *crisis* – of growth or of decline.) In other words, because each of the three systems is a subsystem of the same society, it is linked with the other two. Such connections would be impossible if the systems in question were not concrete systems but either sets of individuals or Platonic or Hegelian totalities.

7.4. THE CULTURAL SYSTEM

In Definition 1 of Section 7.3 the cultural system of a society was characterized as being composed not only of cultural workers (^2P_C) but also of manual workers (^1P_C) and managerial workers (^3P_C). The former are of course directly in charge of cultural production whereas the other two groups are attached to the cultural workers either in an ancillary capacity, as in the case of the maintenance employees of a university, or as organizers, as in the case of university administrators. To take another example, the technicians and typists working for an experimental biologist belong to ^1P_C, the director of the scientist's laboratory belongs to ^3P_C, occasionally to ^2P_C as well.

All members of a cultural system share roughly the same environment and bear certain (social) relations S_C amongst themselves. But not all of the ingredients of a cultural system are on the same footing. Its pivots are: (a) the workers engaged in direct cultural pursuits, i.e. the members of ^2P_C; (b) the part of the natural environment that is the object of contemplation, study, or action, and the part of the artificial environment composed of cultural artifacts such as microscopes, pictures, books, tapes, and files, and (c) cultural work proper – e.g. research, writing, and instructing.

(Ours is of course not the only possible concept of a cultural system. An alternative concept is that including the entire cultural force of the society, i.e. not only the persons working for the sake of culture but also those engaged in an auxiliary capacity by the economic and the political systems, such as the engineers in a quality control unit or the sociologists in a personnel department.

However, this more comprehensive system — i.e. the cultural super-system with composition ^2P — is much less tightly knit or inte-grated. I.e. it is much less of a system than the one characterized previously. Thus the chemical engineer working in an oil refinery is more strongly attached to the economic (and particularly the technical) system than to the cultural system *sensu stricto*. Of course, if he is a research chemist he will belong to both the eco-nomic and the cultural systems and may accordingly be torn by conflicting loyalties.)

Being concrete, a system can be modelled — superficially to be sure — as an input-output box. A possible input-output schema of a culture is a box with inputs of three kinds: work, artifacts, and energy; controls of three types: economic resources, political stimulations and inhibitions, and cultural management; and out-puts of three kinds: cultural artifacts, actions, and waste products. Such a model serves as a reminder that, whether in the narrow or in the comprehensive construal of the notion, a cultural system is a concrete thing rather than "a pattern of behavior", "a collection of beliefs", or "a body of meanings and values". A culture, in fact, is a concrete (material) system embedded in a larger system — a society — and it is composed of living people engaged in activities of various kinds, all involving the neo-cortex of the brain, some of which transcend the biological level and all of which are ultimately social insofar as they involve the entire society.

7.5. STRUCTURE OF A CULTURE

So far we have defined the structure of a society as the collection of all the activities and the interpersonal and interorganizational connections of the components of the society. In short, for any society σ, its structure is $\mathscr{S}(\sigma) = S \cup T$, where S is the set of all social relations and T the set of all the relations of transfor-mation of the environment. The part S of $\mathscr{S}(\sigma)$ is of course the *social structure* of σ. And the subset of S consisting in cultural activities and relations is the *cultural structure* of σ. Thus *the cultural structure is included in the social structure*. Needless to say, like every other structure the cultural structure is the structure of something, in this case the culture, rather than either a concrete thing or a Platonic idea. Cultural structure is a property of culture, which in turn is a concrete system.

The preceding characterization of the cultural structure of a society, though correct, is superficial. (It is like saying that the structure of an algebraic group consists of the group operation and the inversion operation without adding the axioms that define these operations.) A somewhat deeper analysis is as follows.

Consider the various cultural activities in a society. (Recall Section 7.1.) Each of them generates an equivalence relation, such as

> speaking the same language
> holding the same beliefs about nature
> learning the same skills
> listening to the same music
> praying to the same divinities
> observing the same moral rules

and

> playing the same games.

Each of these equivalence relations splits the membership of the society into a number of groups or cells. For example, the relation of speaking the same language splits Canadian society into a number of linguistic groups, the most populous of which are the anglophones and the francophones. And the relation of listening to the same music splits mankind into three main groups – the fans of commercial music, those of folk music, and those of cultivated music.

In general, the ith cultural activity in a society σ generates an equivalence relation $\sim_i$ that induces the ith partition of the membership of M of σ into a certain number of homogeneous or equivalence classes, say m of them. In symbols, the ith partition is the family of sets

$$M/\sim_i = \{M_{1i}, M_{2i}, ..., M_{mi}\}.$$

Every one of the sets in this partition constitutes a *cultural group* (not necessarily a cultural system). And the totality of the cultural groups of a society, resulting from all possible partitions by cultural equivalence relations, constitutes its *cultural structure* (or the structure of its cultural system).

A convenient way of displaying the various cultural groups in the partition induced by the ith cultural equivalence relation $\sim_i$ is to form the column matrix (or, if preferred, vector)

$$C_i(\sigma) = \begin{Vmatrix} C_{1i} \\ C_{2i} \\ \cdot \\ \cdot \\ \cdot \\ C_{mi} \end{Vmatrix}$$

And a most handy way of displaying the entire cultural structure is to collect all the columns resulting from the various partitions, namely thus:

$$C(\sigma) = \begin{Vmatrix} C_{11} & C_{12} & \cdots & C_{1n} \\ C_{21} & C_{22} & \cdots & C_{2n} \\ \cdot & \cdot & \cdots & \cdot \\ C_{m1} & C_{m2} & \cdots & C_{mn} \end{Vmatrix}$$

where n is the number of equivalence relations (hence of partitions) and m is the maximum number of cells or homogeneous groups generated by them. The preceding is of course an $m \times n$ matrix – the *cultural structure matrix*.

Note the following characteristics of the above matrix: (a) all of the entries are sets (of people); (b) some of the entries are likely to be empty, i.e. for some pairs (i, j), $C_{ij} = \emptyset$; (c) all of the non-empty entries are likely to change in the course of time, as new individuals are incorporated into the cells or excluded from them; (d) the algebra of set matrices is specified by stipulating that they add and multiply just like ordinary matrices, with the difference that set union and set intersection appear instead of arithmetical addition and multiplication respectively.

So far our picture of the structure of a culture has been qualitative. If we want numbers, all we have to do is to count the members in each cultural cell C_{ij}. By taking the numerosity or cardinality $|C_{ij}|$ of each cultural cell C_{ij} we get a numerical matrix. An even better picture of the distribution of the membership of a society among its various cultural groups at a given time is obtained by dividing the preceding matrix by the total population at that time. This will show not just the variety of cultural pursuits of a society but also the relative weight of each of them. Thus an elitist culture will be characterized by an extremely uneven distribution of the population among the various cultural groups. More on this in Section 7.7.

What holds for the culture of a society holds also, *mutatis mutandis*, for its economy and its polity. That is, by seizing on 'significant' (real and important) economic relations we can disentangle the economic structure matrix of a society – and similarly for its political structure matrix.

7.6. CULTURAL STRUCTURE INCLUDED IN SOCIAL STRUCTURE

The cultural structure of a society is distinct from both its economic and its political structures. But this difference does not stem from the autonomy of each subsystem. Far from being autonomous, each subsystem of a society interacts with the other subsystems. Thus, as is well known, intense educational activity can modify the economic or the political structure — and conversely. In other words, *the cultural structure is included in the overall social structure* of the society. That this inclusion is literal, not metaphorical, was remarked in the beginning of Section 7.5, where the social structure was conceived of as the set of social relations. The inclusion carries over to the matrix representation of structures, as will be seen presently.

The basic structure of any society is of course its kinship structure. In our schema it is displayed as follows. Take the membership M of the society σ in question and investigate the kinship relations that hold among the members of M. Next build the corresponding kinship equivalence relation, such as having the same mother or, in general, the same matrilineal ancestors. Call $\sim_k$ the kth kinship equivalence relation and partition M by $\sim_k$ to obtain a family of disjoint subsets, i.e. set $M/\sim_k = \{M_{1k}, M_{2k}, \dots M_{1k}\}$ for the kth partition of M. Finally let k run over all the (relevant) kinship relations in σ, to obtain the total *kinship structure* of σ:

$$
\mathcal{K}(\sigma) = \begin{Vmatrix} K_{11} & K_{12} & \dots & K_{1n} \\ K_{21} & K_{22} & \dots & K_{2n} \\ \dots & \dots & \dots & \dots \\ K_{m1} & K_{m2} & \dots & K_{mn} \end{Vmatrix}
$$

The next step is to exhibit the economic structure of the society in question. Suppose we study its economy and discover a number of economic equivalence relations, such as those of having the same occupation, consuming the same kinds of merchandise, wielding the same economic power, and so on. In this way we construct the economic structure matrix $\mathcal{E}$ of the given society. In like manner we analyze the membership of the society into groups of people that exercise the same rights, or have the same political leanings, or wield the same political power, and so on. The outcome is of course the political structure $\mathscr{P}$ of the society concerned.

The four partial structures $\mathscr{K}$, $\mathscr{E}$, $\mathscr{C}$, and $\mathscr{P}$, can be combined into a single matrix, namely the overall structure matrix of the society. This can be done as follows. Take the kinship structure matrix $\mathscr{K}$ and complete it with as many empty set $\emptyset$ entries as needed to accommodate the other three structure matrices. (Take into account that the four matrices need not have the same numbers of rows and columns.) The result will look like this:

$$\mathscr{K}^*(\sigma) = \left\| \begin{array}{ccccccccccccccc} K_{11} & K_{12} & \ldots & K_{1p} & \emptyset & \emptyset & \ldots & \emptyset & \emptyset & \ldots & \emptyset & \emptyset & \ldots & \emptyset \\ K_{21} & K_{22} & \ldots & K_{2p} & \emptyset & \emptyset & \ldots & \emptyset & \emptyset & \ldots & \emptyset & \emptyset & \ldots & \emptyset \\ \multicolumn{15}{c}{\dotfill} \\ K_{q1} & K_{q2} & \ldots & K_{qp} & \emptyset & \emptyset & \ldots & \emptyset & \emptyset & \ldots & \emptyset & \emptyset & \ldots & \emptyset \end{array} \right\|$$

Next proceed similarly with $\mathscr{E}$, $\mathscr{C}$, and $\mathscr{P}$ to obtain $\mathscr{E}^*$, $\mathscr{C}^*$ and $\mathscr{P}^*$ respectively. Finally form the set-theoretic union of the entries of all four starred matrices having the same indices, i.e. compute $K_{ij}^* \cup E_{ij}^* \cup C_{ij}^* \cup P_{ij}^*$ for all pairs (i, j). The resulting matrix, which may be called the dotted sum of the partial matrices, represents the *overall structure* of σ:

$$\mathscr{S}(\sigma) = \mathscr{K}^* \dotplus \mathscr{E}^* \dotplus \mathscr{C}^* \dotplus \mathscr{P}^* \equiv \| K_{ij}^* \cup E_{ij}^* \cup C_{ij}^* \cup P_{ij}^* \|.$$

Imaginary example. If we consider 3 kinship, 2 economic, 4 cultural, and 2 political equivalence relations in a given society, we end up with an overall structure matrix that looks like this:

$$\left\| \begin{array}{ccccccccccc} K_{11} & K_{12} & K_{13} & E_{11} & E_{12} & C_{11} & C_{12} & C_{13} & C_{14} & P_{11} & P_{12} \\ K_{21} & K_{22} & K_{23} & E_{21} & E_{22} & C_{21} & C_{22} & C_{23} & C_{24} & P_{21} & P_{22} \\ \multicolumn{11}{c}{\dotfill} \\ K_{m1} & K_{m2} & K_{m3} & E_{m1} & E_{m2} & C_{m1} & C_{m2} & C_{m3} & C_{m4} & P_{m1} & P_{m2} \end{array} \right\|.$$

where m is the maximal number of rows and where some the entries are likely to be empty, i.e. equal to $\emptyset$.

In short, just as the culture of a society is one of the subsystems of the latter, so its cultural structure (i.e. the structure of the cultural subsystem) is included in the overall structure of the society, i.e. in the social structure of the latter. Similarly with the economy and the polity and their respective structure. (On the other hand the kinship structure is not the structure of a subsystem of society but is the basic structure of the society as a whole.) We can distinguish the three partial structures but we cannot separate them because any change in one of the partial structures is likely to affect the other two, if not immediately then in the long run. But the matter of change deserves a new section.

7.7. STATE AND CHANGE OF A CULTURE

Like every other concrete system, the culture of a society is in a definite state at any given moment. A simple way of describing the momentary state of the culture of a society is to exhibit its cultural structure matrix – which is just a way of stating who is doing, or has done, what cultural activities. If an anonymous and quantitative description is needed, we count the membership of each cultural group and divide the result by the total population N of the society (at the same time). In this way we obtain the *cultural density matrix* for society σ at time t:

$$D^C(\sigma, t) = \frac{1}{N} \, \|\,|C_{ij}|\,\|$$

where the group populations $|C_{ij}|$, as well as the total population N, are taken at time t. $D^C(\sigma, t)$ represents the *state of the culture* of σ at t.

(There are of course alternative representations of the momentary state of a culture. One of them is the exhibition of the instantaneous values of the inputs and outputs of the cultural system, as modelled by the input-output box suggested towards the end of Section 7.4.)

As time goes by the population of each cultural cell may vary. However, changes in the absolute population of each cell may be offset by changes in the total population of the society; i.e. the relative population (or population density) of a cell may remain roughly constant in time. If it does then we speak of a *stagnant culture* — otherwise of a *dynamic culture*. In a dynamic culture some cells grow at the expense of others, subject to the constraint that the sum of the populations of the entries in any given column of the cultural structure matrix equals the total population:

$$\Sigma_i |C_{ij}| = N.$$

Whether a dynamic culture moves forward or backward is another matter. Judgments of progressive or regressive cultural trends hinge upon valuations. Not that they cannot be made, or that they are necessarily subjective: only, they are relative to some value system or other.

The *net cultural change* in society σ between times t_1 and t_2 equals the difference between the corresponding density matrices:

$$\Delta^C(\sigma; t_1, t_2) = D^C(\sigma, t_2) - D^C(\sigma, t_1).$$

Obviously, if a given cell of the change matrix is positive, then it has grown; if it is zero, it has remained stagnant; and if it is negative, it has declined. Because the growth of any cell occurs at the price of the decline of some other cells, the net change in all of the entries has to be watched.

If we wish to trace the history of each cultural cell C_{ij} throughout a given period, all we have to do is to find the sequence of values of the corresponding density D_{ij} over that time interval. In symbols,

$$H_{ij}^C(\sigma; t_1, t_2) = \langle D_{ij}^C(\sigma, t) | t \in [t_1, t_2] \rangle.$$

The total *cultural history* of society σ over the same period is then the full matrix

$$H^C(\sigma; t_1, t_2) = \| H^C_{ij}(\sigma; t_1, t_2) \|.$$

The preceding representation of cultural states and trajectories provides a framework for a phenomenological (i.e. superficial) account of cultures and culture changes. Such an account over-looks the internal dynamics of cultural production and diffusion as well as the interactions between the culture on the one hand and the economy and the polity on the other. These interactions can be accounted for by a sort of Leontieff input-output analysis of the main subsystems of the society concerned. Indeed one can, at least in principle, set up the *total* activity of the society at a given moment or over a given time period:

$$A = \| A_{mn} \|, \quad \text{with} \quad m, n = 1, 2, 3,$$

where

A_{11} = The part of the cultural output that stays in the cultural system (e.g. poetry).

A_{12} = The part of the cultural output that is absorbed by the economy (e.g. applied research).

A_{13} = The part of the cultural output that is absorbed by the polity (e.g. political philosophy).

A_{21} = The part of the economic output spent in cultural pursuits (e.g. laboratory equipment).

A_{22} = The part of the economic output that is reinvested in the economy (e.g. tool machines).

A_{23} = The part of the economic output that is absorbed by the polity (e.g. government budget).

A_{31} = The part of the political activities devoted to controlling the culture.

A_{32} = The part of the political activities aimed at controlling the economy.

A_{33} = The part of the political activities devoted to maintaining the political system.

Each entry of the total activity matrix is an extremely heterogeneous set composed of people and nonhuman things, of human activities and physical processes. Some of the subsets included in each entry can be assigned numbers, as in the case of numbers of people and of work hours, electric power and the price of commodities. But whether or not a certain entry is quantifiable, fully or in part, in principle it is possible to set up models for the evolution of the total activity matrix. As a matter of fact this is what cultural, economic, and political planning are all about. Only, such planning is usually partial, hence lopsided, rather than integral, insofar as it concerns just selected aspects of the total activity matrix. No wonder they fail more often than not. Only global models and global plans can work, precisely because of the interrelations among the various subsystems of any given society — interrelations exhibited though not explained by the total activity matrix. But enough of this, for our sole purpose in bringing in this matrix was to emphasize the thesis that culture, though distinguishable from both economy and polity, is not detachable from them.

7.8. THE CULTURE OF MODERN SOCIETIES

The cultures of primitive societies are rather monolithic inasmuch as they are not composed of subsystems. So are their economies and polities. The emergence of agriculture, and later on of civilization, were accompanied by an unprecedented division of labor, in particular of cultural work. The single shaman or medicine man was succeeded by a cohort of priests, healers, bards, teachers, painters, and later on scribes and experts of various descriptions. Eventually the cultural system became split into a number of subsystems — the religious system centered around temples, the educational system centered around schools, and so on. These various subsystems complemented one another in certain respects but were at odds in others, if only because they had different goals and competed for finite human and natural resources. In any case they interacted, and this interaction was a source of their evolution.

Because a culture is a subsystem of a society it has its own dynamics — hence some measure of autonomy — and it also interacts with the remaining chief subsystems of the society, namely its economy and its polity. Culture, in sum, is neither totally free and omnipotent nor wholly enslaved and powerless. Just as some members of the economic system exercise economic power, and some members of the political system wield political power, so some members of the cultural system wield cultural power, particularly if they are entrenched in certain cultural subsystems, whether governmental or private. For example, the school system of a region exerts some influence on all the inhabitants of the region — often as strong an influence as that exerted by organized religion in the past. This influence is not restricted to purely cultural matters. Thus a creative cultural organization may study, discuss, and even propose and advertise blueprints for economic and political action. Such proposals do not amount to economic or political action but they may arouse action and guide it — for better or worse. After all, people are being moved by ideas all the time — or rather by people with ideas.

A flourishing culture is one that teems with novelty — enlightening or enjoyable novelty — and is free to pick the best of it without too much destructive interference from the economic and the political subsystems. A declining culture is one that has ceased to value discovery and invention, taking refuge in routines, in lamaistic repetition, or in withdrawal from reality. The management of a cultural system — the politics and economics of culture — can encourage creativity, channel it, or kill it. A "dead" culture is still a culture and one that can remain in this state for a long time. But, because of the functional relationships among the three main subsystems of every society, any major change in the economy or the polity is bound to have cultural repercussions. And conversely, any major cultural achievement is bound to have an impact on the economy or the polity, particularly in modern society.

The cultural system *coevolves* then with the economic and the political systems. However, the modes of evolution of these subsystems of society may be quite different. In fact, cultural growth may be compatible up to a point with zero economic growth and political stagnation — provided the political system does not exert too strong a distorting influence on cultural production. Moreover, while economic development is limited by natural resources, there are no similar limits to cultural growth. This holds at least for the cognitive aspect of cultural evolution: the more we know the more new problems we can pose and solve. This is not to deny that there are economic and political constraints on cultural evolution. For one thing any given society can support only so many professional composers, mathematicians, or historians. Yet in the case of the culture, unlike that of the economy, there is a solution in sight, namely automation combined with the amateur pursuit of cultural interests.

7.9. CONCLUDING REMARKS

We conceive of a culture as a subsystem of a society, which is in turn a concrete system — as concrete as a stone or an organism, a forest or a flock of birds. We assume that every society has a culture, an economy, and a polity — however primitive each of these may be — and that all of these systems interact pairwise. Moreover we conceive of the structure of each subsystem of a society as included in the overall social structure. In particular the cultural structure of a society is part of the social structure of the latter. Furthermore we conceive of history — whether cultural, economic, or political — as the evolution of social structure or, equivalently, as the sequence of states of the social system. Hence in our view *culture change is one component of social change* — the other two components being of course economic change and political change.

The systems view of culture sketched in this chapter is at variance with cultural idealism, according to which "Culture is but a body of ideas and values". There is no such thing as a disembodied value: there are only concrete people who think up ideas and evaluate. Our concept of culture is also at odds with a watered-down version of the idealist conception that equates culture with a set of behavior patterns, and so on. We find it impossible to detach the patterns of feeling, thinking, evaluating, and acting, from the feelings, thoughts, evaluations, and actions of real people. There is no such thing as a behavior rule separate from the corresponding ruled behavior. There are only people behaving in certain ways.

Our concept of culture is also at variance with vulgar materialism, whether in its biological (or environmentalist) version — "Culture is but the way humans adapt themselves to their natural environment" — or in its economic version — "Culture is but an epiphenomenon of the economy". While every cultural system is embedded in a natural environment, the latter is not omnipotent, as proved by the fact that one and the same environment can support in succession rather different societies, hence cultures, just as a given culture can survive, within bounds, in different environments. As for the economy, while no culture can survive without a viable economy, no economy can meet drastic environmental and social challenges without the assistance of a creative culture. There is no prime mover in society: any of the three main subsystems may initiate an important social change, and every such change is likely to have three components — economic, cultural, and political.

Our systems view of culture is also incompatible with the dualist schema of society as formed by an ideal superstructure mounted on a material infrastructure. In our view a society is a thoroughly concrete thing and so is every one of its subsystems. In particular a culture is composed of living persons interacting with one another and with nonhuman components, both natural and artificial; a superstructure, on the other hand, is supposed to be a set of ideas, values and norms — just like the idealist conception of culture.

On the other hand the conception of culture sketched in this chapter agrees with materialist ontologies. It also agrees with the general systems approach and moreover suggests dynamical models of culture. And it is consistent with the growing suspicion among sociobiologists and ethologists that several animal societies other than ours have their distinctive if primitive cultures. Last, but not least, the present systemic view of society fits in with the global approach anthropologists are justly proud of: indeed, even when they talk as if a society were a culture, they in fact treat culture, polity and economy as so many interdependent subsystems of society.

Does our materialist and systemic view of culture imply that languages, techniques, games, musical scores, poems, mathematical theorems, scientific theories, philosophical views and other cultural items are material things? Not at all: the implication is that such objects have no existence by themselves, i.e. detached from their creators or users. In isolation from the latter, such items either do not exist at all or they are mere chunks of matter (e.g. sculptures and books). What does exist as a cultural object is not a poem (or game or symphony or theorem) by itself, but the activity of writing, reading, or reciting a poem. Better: What does exist is a person performing such activity in her environment, or a system of persons held together by such activity − e.g. a string quartet, a chemistry class, or an archaeological excavation.

In other words, just as a left parenthesis is not a significant symbol (but a syncategorematic sign), so what are usually called 'cultural objects', such as ideas, moral rules, and records are only components of genuine cultural objects. We do not render them more real, alive, or powerful by taking their totality (set) and calling such set 'the world of culture', for sets are concepts. What are real are the persons, and the systems of persons, performing cultural activities with the help of sundry cultural tools such as pencils, microscopes, calculators, violins, or chisels. Eliminate such individuals or deprive them of such tools, or dissolve the cultural systems they compose, and the entire culture will be damaged or even destroyed.

A culture, like an economy or a polity, is a concrete system and as such one characterized by a composition, an environment, and a structure (which includes cultural activities and relationships). It is true that a correct description of the environment and the structure of a culture calls for certain nonphysical features such as information flows, goals, intentions, decisions, values, opinions, and tastes. However, this only shows that physicalism, or vulgar materialism, is inadequate: that it must be expanded to accomodate supraphysical (but not ghostly) properties and relations. Emergentist systemic materialism (Chapter 2) does make room for such properties, so it is competent to account for culture.

Some social scientists might be tempted to ask: So what? What influence could such a materialist view of culture exert on social science research? Is not science characterized by a method rather than by an ontology? My reply is that a method is only one component of an approach, the others being a general framework, a set of problems, and a set of goals. The general framework is a body of philosophical (logical, ontological, epistemological, and ethical) principles. These principles are not decorative: they guide, misguide, or block the cognitive enterprise from the formulation of problems to the evaluation of solutions. In the case of the scientific approach, its overriding ontological principle is that reality (the world) is composed only of concrete (material) things, in particular systems, in lawful flux. (For the ontology of science see Bunge, 1977a.) Clearly, this principle guides scientific research, by enjoining us to investigate only concrete things (never disembodied items), their interactions, and their laws. Our conception of culture as a material (yet supraphysical) system complies with this principle. And it saves us from the mistake of conceiving of culture as a self-existent ghostly entity. But this mistake deserves a chapter of its own.

CHAPTER 8

POPPER'S UNWORLDLY WORLD 3

At the Third International Congress of Logic, Methodology and Philosophy of Science, held in Amsterdam in August, 1967, Sir Karl Popper astonished the philosophical community by formulating his doctrine on "the third world", or world of ideas, or "objective mind", which he later called 'World 3'. In his paper, titled "Epistemology without a knowing subject", he wrote as follows: "without taking the words 'world' or 'universe' too seriously, we may distinguish the following three worlds or universes: first, that world of physical objects or of physical states; secondly, the world of states of consciousness or of mental states, or perhaps of behavioral dispositions to act; and thirdly, the world of *objective contents of thought*, especially of scientific and poetic thoughts and of works of art" (Popper, 1968, p. 333).

This thesis came as a surprise, not because it was new — which it was not — but because until then Popper had been a merciless critic of idealism. In particular he had criticized the objective idealism of Plato and Hegel (Popper, 1945) as well as the subjective idealism of Berkeley (Popper, 1953), particularly for having influenced contemporary positivism. And now, without warning, Popper makes a right-about turn, or so it seems, and adopts an idealistic stand. He owns the latter explicitly: "what I call 'the third world' has admittedly much in common with Plato's theory of forms or ideas, and therefore also with Hegel's objective spirit, though my theory differs radically, in some decisive aspects, from Plato's and Hegel's. It has more in common still with Bolzano's theory of a universe of propositions in themselves and of truths in themselves, though it differs from Bolzano also. My third world

137

resembles most closely the universe of Frege's objective contents of thought" (Popper, 1968, p. 333). In his *Autobiography* (in Schilpp, 1974) Popper has confirmed this philosophical self-analysis.

8.1. A CONVERSION?

How is this sudden conversion of Popper's to objective idealism to be explained? He himself does not explain it in his *Autobiography*. He tells us only that, just like Bolzano, he had wondered for long years about the ontological status of "propositions in themselves". He writes also that he did not publish anything on the third world until he "arrived at the conclusion that its inmates were real; indeed, more or less as real as tables and chairs" (Popper, 1974, p. 146).

It would seem that Popper suffered a late conversion from the all-round anti-Platonism of *The Open Society* (1945) but he himself holds that this was not the case. In a letter of October 4th, 1977, Sir Karl informed me that "The Amsterdam paper of 1967 (published in 1968) was a re-written paper first read to my LSE [London School of Economics] Seminar in 1959 or 1960. The ideas of this paper go back right to the *Logik der Forschung*, and to the *Wahrheitsbegriff* of Tarski's [1935/36]. Also, the *difference* between my views and those of Plato and Hegel etc. is very great; even that between my views and those of Bolzano and Frege. In the Amsterdam paper, I stress similarities more than differences. Main difference: world 3 is the *product* of human minds. (But there is a strong feedback.) World 3 can act upon the physical world 1 (although only in an indirect way, through world 2)."

However, the doctrine of knowledge without a knowing subject, which is part of the World 3 doctrine, is a direct generalization of the thesis of the objectivity of scientific knowledge, maintained by Popper himself with regard to the quantum theory, the same year he wrote his Amsterdam paper (Popper, 1967). Besides, Sir Karl's letter poses the following problem to the historian of philosophy. If the doctrine was found *in nuce* both in the *Logik der Forschung* and in Tarski's equally famous paper, why did nobody seem to notice it in the course of three decades? Why has Tarski always regarded himself as a materialist and, more particularly, a nominalist? (Tarski, personal communication, Jerusalem, 1964.) And why did Popper start his Amsterdam lecture by stating: "I shall make an attempt to challenge you, and, if possible, to provoke you"? (Popper, 1968, p. 333).

I suspect that what Tarski's paper on (formal) truth and the *Logik der Forschung* did contain is something very different from Plato's ideas and Hegel's objective spirit. (These two expressions occur in Popper's 1968 article as synonyms of 'the third world'. The synonymy is repeated next year in the lecture at the International Congress of Philosophy, entitled "On the theory of the objective mind".) What we do find here is a tacit rejection of psychologism, i.e. the doctrine according to which propositions are thoughts, and rules of inference laws of thought. We also find in these works a sort of formal objectivism consisting in the thesis that, once the rules of the game have been accepted, there is no arbitrariness, for everything proceeds according to rule. To be sure, both the logician and the mathematician treat propositions and inference rules *as if* they enjoyed an autonomous existence. But this may make them fictionists, not necessarily metaphysical (Platonic) realists.

Be that as it may, the fact is that Popper has proposed the partition of the world into three, and that this doctrine revolves around psychophysical dualism. Let us then take a look at the latter.

8.2. MIND–BODY DUALISM IN POPPER'S PHILOSOPHY

Popper has become a staunch defender of psychophysical dualism. Jointly with his friend, the eminent neuroscientist Sir John Eccles,

Popper has been looking for arguments in favor of the ancient thesis that the mind is an entity or substance separate from the body though interacting with the latter. In particular, the Popper and Eccles volume *The Self and Its Brain* (1977) expounds the thesis that every one of us is an embodied mind. (Actually Popper defends two different theses in that book, although he does not seem to realize that they are different: one is interactionism, the other is Plato's doctrine that the mind steers the body as the helmsman steers the ship.)

Psychophysical dualism is, of course, the most popular doctrine about mind, at least in the West. We adopt it tacitly in daily life when we speak of the influence of ideas on bodily states and behavior, and conversely. We espouse it when saying that a given state or process is psychological, not physiological, and when we distinguish a thought process from its "products". Dualism is inherent in psychoanalysis and spiritism, in Platonism and Cartesianism, in Christianity, Islam, and Buddhism. The doctrine is so entrenched, and it has been defended so zealously by conservative ideologies and the corresponding institutions, that we hardly realize it. In particular, the neuroscientist who writes about 'the neurophysiological correlates' of mental states, without explaining what the latter are nor how they are 'correlated' with their material counterparts, does not realize that he is the prisoner of a vulgar ideology.

This is not the place to perform a critical analysis of psychophysical dualism, a task done elsewhere (Bunge, 1980). We must restrict ourselves here to stating dogmatically that the dualist describes the mental in vulgar terms, not in scientific ones, and that he resists any identification of mental states with brain states by resorting to equally vulgar arguments, such as the difference between the language of introspection and that of physiological psychology, and the difference between concepts and brain processes.

What we must point out here is the *vagueness and irrefutability* of psychophysical dualism, because both characteristics ought to have made it unacceptable to Popper. That psychophysical dualism is as imprecise as it is untestable, seems clear from the following.

(i) Whereas the scientist takes it for granted that *every state is a state of some concrete entity* (physical, chemical, biological, social, or what have you), the dualist talks about mental states *in themselves*. (He may no longer dare to speak of states of a *res cogitans*, or mental substance, since anybody could ask him for his reasons for calling 'substance' that which lacks substantiality.) That this is a case both of reification and of confusion between distinction and detachment seems clear. It is also clear that dualism consecrates the traditional detachment of psychology and psychiatry from neuroscience.

(ii) The conceptual imprecision of psychophysical dualism is such that it lacks a *theory* proper. It confines itself to giving some *examples* of mental activity and to saying what is *not* mental, and in particular to insisting that the mental is not physical — which is obviously insufficient since life too is not merely physical yet it is nonmental. Psychophysical dualism is so imprecise that it has not been mathematized.

(iii) Being imprecise, psychophysical dualism is *untestable* and, in particular, irrefutable, for it cannot issue precise predictions that can be put to the test. The fact that when in an emotional state one does not see well and does not coordinate one's movements well, so that one cannot drive correctly, can be explained in two ways. The dualist explanation is simple and therefore popular: your mind is acting upon your body — in a mysterious way, perhaps incomprehensibly. (Eccles [1951] suggested psychokinesis.) The psychophysiological explanation is more complex: it resorts, in particular, to the action of certain hormones on the synapses of the neurons of the motor system. By experimentally varying the concentration of such hormones one may control, at least in principle, the "somatic effect" of the "emotional state of mind" (as the dualist would put it). The interactionist dualist will tell us that this is nothing but an example of the action of body on mind — which he has never denied. And if the hormone injection were not to change perceptibly the subject's behavior, the dualist may argue that in some cases the mental states or processes are so intense that no neurophysiological process can alter them. He may cite the experiences of the martyrs who died singing and of the yogis. There is no way of refuting those who hold an irrefutable doctrine — unless of course they realize that such a doctrine does not explain anything nor, consequently, contribute to posing or solving any scientific problems.

In conclusion, psychophysical dualism is inimical to Popper's own methodology. However, this contradiction is not as important as the possibility that it be one of the sources of his World 3 doctrine. And, no matter what the genesis of this doctrine may have been, we must examine it to find out whether it is true. Let us do this.

8.3. THE PLURALITY OF WORLDS

At the very beginning of his first paper on the tripartition of the world, Popper warns us that we must not take "too seriously" the words 'world' and 'universe', which occur in the formulation of his theory. (Incidentally it is a thesis not a theory.) However, we cannot take that warning seriously: if we are asked to take seriously a given thesis, then we must take just as seriously the key terms occurring in its formulation. It won't do to tell us — as Popper has done repeatedly — that matters of words and their meanings are insignificant. They may be so in inexact philosophy, but they are just as important in exact philosophy as they are in mathematics and science.

Strictly speaking, *the world* (or universe) is the supreme concrete thing, i.e. that thing which contains (as parts) all other concrete things. This is the way physical scientists use the word. In a figurative sense, *a world* (or universe) is either a subsystem of the universe (such as our planet) or a structured set of objects, whether concrete or conceptual. Taken in its strict sense, the denotatum of 'the world' is concrete and, of course, unique. On the other hand, in a figurative sense there may be as many "worlds" as concrete systems and conceptual systems. The solar system is one of them and the set of natural numbers is another. Likewise we may talk of the "world" of the fish of a given species, or a given region, as well as the "world" of philosophical ideas.

There are many such partial "worlds" and not all of them are concrete or material. Undoubtedly, it may be convenient to use a single expression of the form *the world X* (or *the X world*) to denote some concrete system (such as an ecosystem) or a structured set (such as a topological space). However, verbal economy must often be paid for with conceptual confusion. This is exactly what happens with the term 'world' in Popper's thesis of the trinity: in it the word 'world' is not used uniformly, i.e. with a single definite signification. Indeed let us recall how Popper defines his "worlds".

World 1, or the physical world, is "the world of physical objects or of physical states". This is an ambiguous phrase. In fact such a "world" may be a concrete individual (namely the system composed of all material things) or a set and therefore a concept. (If the components of "world 1" are physical states, then this "world" is a set and therefore an *être de raison*.) Clearly, any statement about "world 1" will depend critically on whether it is taken as a thing or as a set. In the first case it may be attributed physical properties, not so in the second.

World 2, or the psychical world, is "the world of states of consciousness or of mental states, or perhaps of behavioral dispositions to act". It would seem that apes and, a fortiori, other higher vertebrates have no access to this "world", which is rather odd given the available evidence on their mental life and given the theory of evolution. It would also seem that this time the world is a set — unless the states of consciousness of one person are allowed to influence directly those of others. (Needless to say, to a materialist the elements of such a set are ghostly, for one can speak properly of brain states but not of states of consciousness as distinct entities.) Moreover it would seem that, while world 1 is eternal (at least in the forward direction of time), world 2 is not so necessarily. Indeed if no thinking beings were to remain, world 1 would continue to tick but world 2 would become extinct — though presumably leaving world 3 behind.

World 3, or the cultural world, is "the *objective contents of thought*, especially of scientific and poetic thoughts and of works of art". Among the "inmates" of this "world" Popper lists problems, critical arguments, and theories, as well as the "contents" of books, journals, and libraries. Sometimes the books, journals and other material embodiments of intellectual and artistic work are included as well. Thus: "I regard books and journals and letters as typically third-world objects, especially if they develop or discuss a theory" (Popper, 1974, p. 145). There are thus both *embodied* World 3 objects, such as gramophone records, and *unembodied* World 3 objects, such as numbers. (E.g. Popper and Eccles, 1977, p. 41.) In short, World 3 is composed of all the "products" of mental activity, or World 2 inmates. And that "world", like World 2, is a set. But unlike the members of World 2, which are perishable, some of the inmates of World 3 are, or at least are very close to being, eternal objects in the manner of Plato, Hegel, Bolzano, Frege, Husserl, and Whitehead. Let us take a closer look at this idea.

To clarify his idea about World 3, and perhaps persuade us that it exists by itself once created by man, Popper imagines two situations which many have envisaged since the beginning of the nuclear bomb age. In the first a world-wide blast destroys all of our cultural artifacts except for the libraries and museums: these remain, and our ability to learn from them is left intact. It is certain, says Popper, that after long sufferings "our world" (in this case industrial civilization) may again be set in motion. (That this is doubtful rather than certain makes no difference to the philosophical argument.) In the second mental experiment there remain some human beings but all the libraries and museums are destroyed by nuclear bombs. In this case our ability to learn from books and journals would be useless, and it would take millenia to reconstruct civilization. (Why anybody would wish to reconstruct a civilization capable of self-destruction once in a while is not explained, but is not relevant to our problem either.)

The above "experiments" exhibit "the reality, significance and the degree of autonomy of the third world (as well as its effects on the second and first worlds)" (Popper, 1968, p. 335). Popper is satisfied with such *Gedankenexperimente* even though in his *Logic of Scientific Discovery* he had criticized, and rightly so, the physicists who claimed to prove theorems by imagining experiments. Others won't remain satisfied with a couple of science fiction stories, anymore than they are persuaded by Walt Disney's movies that Mickey Mouse is real. Let us see why.

8.4. CRITICISM OF THE FANTASY

Only a crass materialist would dare deny the importance of ideas — or, rather, that of thinking brains. But this does not entail that ideation may constitute a world ("world 2") nor that the "products" of ideation (its "contents") constitute a "third world" enjoying autonomous existence from the moment of its coming into being. The least that can be objected to the thesis of the real

existence of World 3 is that it is imprecise; the most, that it is
groundless. Let us see why.

(i) Firstly, Popper does not tell us clearly what he means by
'real' and 'reality'. For one thing he does not seem to regard the
predicate "is real" as dichotomous: in fact he tells us that the
inmates of World 3 are "more or less as real as tables and chairs"
(Popper, 1974, p. 146). To most other philosophers — the excep-
tions being some Thomists such as Jacques Maritain — reality does
not come in degrees: every object is either real or unreal. Also,
many philosophers are careful to distinguish two concepts of
existence or reality: material and conceptual, and to state that,
whereas some objects exist materially, others exist conceptually.
(More in Chapter 9.) Finally, all philosophers interested in' the
problem of reality have attempted to clarify this concept. In particular, materialists equate "real" with "material", and in turn
identify "material" with "changeable". (See Chapter 2.) Popper
leaves us in the lurch on this point.

(ii) In the second place, it should be necessary that all the
inmates of World 3 be ideal objects, and preferably also of the
same kind (e.g. sets). The reason is that ideal objects do not combine with material ones, forming mixed systems. Material objects
can combine with one another to produce material systems, and
ideal objects can associate, forming ideal systems. There are no
mixed entities except in hylomorphic metaphysics, which Popper
does not approve of explicitly. Moreover if one holds that there
are such mixed beings then one is supposed to exhibit, if not
empirical evidence for or against such a hypothesis, then at least a
calculus containing the operations producing them out of material
entities on the one hand and ideal objects on the other. (It won't
do to say that a written sentence is the *embodiment* of a proposition in itself, and that a theorem is an *unembodied* object: these
are vague words suggested by religion and psychophysical dualism.
Give me a calculus of embodiment and unembodiment and I may

begin to take you seriously.) In short, World 3 does not constitute a world proper (i.e. a system) as long as its ideal inmates are allowed to mix promiscuously with the material ones without obeying any laws.

(iii) Thirdly, Popper does not explain what he means by "the content" of a drawing, a musical phrase, or any other nonconceptual component of World 3. Are they "messages", and if so how about abstract art? Neither he nor anyone else seems to have proposed a semantic theory equally applicable to works of art and scientific theories. Because of this, no less than (i) and (ii), the World 3 thesis is inexact to the point of meaninglessness.

(iv) In common with the idealist metaphysicians who preceded him, Popper starts by considering the intellectual and artistic activities of human beings, and ends up by abstracting from such concrete entities and their activities, to focus his attention on the "products" (and sometimes also their "embodiments", i.e. books, paintings, phonograph records, etc.) into a single "universe", namely World 3. I.e. he assumes that this heterogeneous set constitutes a system. Finally, forgetting how this pseudosystem came about, he declares that it leads an autonomous existence, i.e. one independent of its creators – and this simply on the strength of two *Gedankenexperimente*. What is this but *reification* (thingation) together with *systematization by decree*? What is it but to take literally the fire-ashes metaphor, or the model of the factory and its products?

(v) Popper asks us to attribute autonomy to World 3, i.e. to assign it an existence independent from its creators and probably also from everything else. From everything? If there were a Supreme Creator and Annihilator capable of annihilating every material thing, to the last electron, photon and neutrino, would World 3 subsist? Being an agnostic, Popper has not envisaged this third *Gedankenexperiment*, so we won't know for sure what the ontological status of World 3 is.

(vi) Popper does not justify or corroborate, let alone try to refute, the conjecture that his World 3 exists or is real. Why does he think that his readers, usually attracted by his critical rationalism, should be able to swallow this monster of traditional metaphysics — the monster that gave metaphysics its bad name among scientists?

8.5. KNOWLEDGE: SUBJECTIVE AND OBJECTIVE

It is well known that the Greeks distinguished between *doxa* (opinion or uncertain or subjective knowledge) and *episteme* (science or certain and objective knowledge). Popper keeps this distinction though leaving certainty aside: to him all knowledge, even mathematical knowledge, is conjectural and therefore uncertain and subject to revision.

Still, though human knowledge is fallible and therefore subject to revision, at least it can be objective: it need not be subject-dependent. (Subjective knowledge, or mere belief, does not deserve to be called knowledge, Popper tells us in 1972, Chapter 2.) Unfortunately Popper does not define what he means by 'objective knowledge'. (In general he refuses to define his terms, alleging that definitions lead nowhere.) However, from the context it would seem that Popper calls 'objective knowledge' all knowledge that does not depend upon the knowing subject — although without saying whether the independence is referential or alethic. (A proposition can be said to be *referentially* subject-independent if, and only if, it does not concern any particular subject; and *alethically* subject-independent if its truth value is the same for all knowing subjects.) I shall argue that, if this were so, there would be no knowledge.

To the psychologist knowing is a mental (or brain) state or process of some animal. The same to an epistemologist other than Popper. When we say '*p* is known', where '*p*' designates a proposition or denotes a fact, we do nothing but abbreviate 'There is at least one animal that knows p'. (Actually we intend to say that there are several animals that know p, or that whoever is not an ignoramus knows p, or that anyone may get to know p if only he intends to. But this is unimportant in our case.) Similarly, when we assert that q is unknown, we abbreviate 'No animal knows q', or at least 'None of the animals I know knows q'.

All knowledge is then knowledge of *something by somebody*, whether human or not. In particular, that somebody may be you or I. If either of us asserts "I know p", and this statement proves to be true (e.g. as a result of a test), "it is" concluded − i.e. anyone can validly conclude − that there is at least one animal that knows p, i.e. that "p is known (by somebody)".

At first sight, whereas 'I know p' is a subjective expression, "p is known" is a statement of objective knowledge. But "I know p" implies "p is known" (i.e. "There is at least one animal that knows p"). Hence the result would be that subjective knowledge not only originates objective knowledge but is also its foundation. However, this is not so, for "I know p" may be true or false quite aside from my certainty or uncertainty concerning the propositions "p" and "I know p". The same holds for "p is known": this "objective" proposition may be empirically true or false (to some extent). And this is what matters most in both cases: the degree of truth of the proposition p.

Neither the mathematician nor the physicist nor the sociologist are in the habit of writing sentences of the type 'p knows q', except when investigating empirically what people know (e.g. when grading exam papers). They do publish, on the other hand, sentences of the forms 'p', 'not-p', 'If p then q', etc., which do not refer explicitly to those who formulate or believe them. Even when the social psychologist or the historian of ideas make statements about beliefs, they do so without putting themselves in their sentences. Thus they will write 'People of kind X tend to believe p', instead of 'I tend to believe that people of kind X tend to believe p'.

In this sense the data and the hypotheses of science, formal or factual, basic or applied, *are* objective, or referentially and alethically free from the knowing subject. When stating that the statements of science are objective we do not mean that they exist by themselves as inmates of World 3. All one means is (a) that what matters is the referent rather than the speaker or writer, and (b) that the proposition in question has to be judged according to canons accepted beforehand rather than by its agreement or disagreement with some authority.

In other words, that a cognitive statement is objective does not entail that, once made, it has a "life" of its own or that it enters World 3. At most one *feigns* that the proposition exists by itself, and this simply because it can be thought and examined by others similarly (but not identically) to the way that the vegetables and fruits in an open-air market can be freely examined by the public. That fiction is indispensable if we wish to divert attention from the subject to what he asserts or the way of justifying (proving or confirming) or else refuting the proposition concerned. It is also indispensable to deal with infinite sets of propositions (such as theories), since no finite being could think up all of them. (More on fictionism in Chapter 9.)

8.6. TWO APPARENT EXCEPTIONS

There are, however, two fields of scientific research that are sometimes believed to have been invaded by subjectivism, namely the quantum theory and psychology. It can be shown that both beliefs are false. The subjective interpretation of the quantum theory can be shot down with heuristic arguments (e.g. Bunge, 1955; Popper, 1967) or rigorously (Bunge, 1967, 1973). What one does in the latter case is to analyze the "variables" (functions and operators) of the theory and to organize the latter axiomatically. The analysis fails to exhibit any references to a knowing subject, such as an observer. And the axiomatization fails to reveal any assumption about experimenters. To put it in a positive fashion: every formula of the quantum theory refers exclusively to physical entities, none to knowing subjects.

What happens is that it is easy to "interpret" any proposition p as "There is an observer that verifies p". For example, instead of saying that properties P and Q are related by some function F such that $Q = F(P)$, one can (but one ought not to) say "The knowledge of P determines that of Q when the computation indicated by '$Q = F(P)$' is performed". This is just a didactic (but misleading) prop that does not prove that the function F is subjective, in the sense that it is left to the arbitrary decision of the knowing subject.

Conversely, every pragmatic expression, such as "The values an observer may obtain when measuring the dynamical variable P are the eigenvalues of the operator representing P" may (and must) be translated into an expression free from any knowing subject, such as "The P values that a physical entity may take on are the eigenvalues of the operator representative of P". The reason for preferring the latter reading of the formulas of the quantum theory is that it deals with physical problems, not with psychophysical ones. So much for the pseudosubjectivity of the quantum theory.

What happens with psychology is this. Here the referents are (experimental) subjects, which are often capable of knowing. Moreover, the psychologist can be his own subject of observation or experiment. But those who study such subjects scientifically are supposed to proceed objectively: their results are supposed to refer to entire classes (e.g. species, occupations, or age groups) and they are supposed to be publicly scrutinizable. This does not exclude introspection (whatever that may be) but renders its data mere heuristic starting points. The psychologist does not use such data as an unshakable foundation but as a source of hypotheses to be tested objectively.

The objectivity of psychology does not consist, then, in that it is not interested in subjective experience, but in that its conjectures, data, and conclusions are, it is hoped, true irrespective of who formulates them. (Thus we do not believe in the hypothesis of the stages in biopsychological development just because it was formulated by Jean Piaget but because it has been confirmed by many other psychologists and agrees with what neurophysiology tells us about the maturation and plasticity of the central nervous system. On the other hand we do not believe in projective tests, such as the Rorschach, because the "interpretation" of the ink blots is left to the imagination of the psychologist — so that they rarely work.) In short, scientific psychology is no less objective than the other factual sciences.

8.7. EPISTEMOLOGY WITHOUT THE KNOWING SUBJECT?

The last two sections can be summarized as follows:

(i) *All knowledge is knowledge of something by somebody.* There is no knowledge of nothingness or knowledge without a knowing subject. (But of course the object or referent of knowledge can be imaginary, as is the case with those scientific theories that postulate entities that are eventually recognized as unreal.)

(ii) *Objective knowledge is intersubjective and (partially) true knowledge* — i.e. it is invariant with regard to subject replacement even though it must have been discovered or invented by some subject to begin with.

In other words, an item of knowledge is objective not because it exists or subsists in a separate "world", one above corruption (to employ Platonic language) or sheltered from nuclear bombs (to use contemporary language). It is objective because, and to the extent that, there are animals capable of acquiring it and putting it to the test with the help of criteria independent of personal factors such as authority or firmness of conviction. The degree of objectivity of a proposition may be estimated by virtue of the rules of the knowledge game, such as those of logic and empirical testing.

Now, epistemology is concerned with knowledge in general and, in particular, scientific knowledge. And scientific knowledge, unlike mere opinion − which can be subjective and groundless − is objective or invariant with respect to the knowing subject. In other words, the rules of science, though not eternal, are not arbitrary. (Hence scientific research is not really a game.) And although such rules are proposed, discussed, applied, violated, or modified by living researchers, they are not proclaimed on the strength of such personal considerations, for they are supposed to lead to truth, in particular factual truth.

However, the objectivity of scientific knowledge does not imply that it is above all knowing subjects nor, in particular, that it constitutes a "world" independent of its creators (and destroyers). Getting to know something is, by definition, a process or activity and, like every other activity, it can be formalized as a relation that is at least binary. We think and say '(Subject) X knows (fact or proposition) p', not 'X knows' nor just 'p', when we deal with the process of knowledge, i.e. when we consider knowledge as a biological, psychological, or cultural fact. It is only when we are concerned with the outcome or "product" of this process that we disregard the subject and focus on "the content of knowledge", such as a proposition or a theory. But when we do so we engage in something other than epistemology.

Epistemology cannot dispense with the knowing subject because, by definition, it is concerned with what the subject can know, how he gets to know, and related questions. For example, genetics is concerned with genotypes and their relations with phenotypes. On the other hand the epistemologist is interested in the ways the geneticist investigates his genotypes and phenotypes, what motivates him, what guides or misguides him, what he succeeds in discovering or inventing, what are the philosophical presuppositions of his research, etc. To be sure, some epistemologists are not interested in real knowledge but speculate on abstract knowledge by an ideal knowing subject. They are anti-historical, anti-sociological and anti-psychological epistemologists who deal with the fiction of "objective knowledge without a knowing subject".

Whereas the classical epistemologist dealt with the knowing subject and his activities in general, some contemporary epistemologists have understood that the suprahistorical knowing subject hovering above society is a fiction. We have come to know that every knowing subject is a member of a given culture, and that membership in a culture opens up some horizons while it may close others. Contemporary epistemology does not ignore the history of knowledge: it takes the "context of discovery" as well as the "context of justification" into account. (Which is not to say that there is never objective truth, that truth value assignments are only seals of social approval.) In particular, the new epistemology that is in the making studies the factors of various kinds (cultural tradition, economic potential, political régime) that stimulate or inhibit the production and circulation of knowledge. In particular, epistemology must take into account that scientific research is just one cultural activity, hence the study of it cannot be isolated from the study of other branches, in particular philosophy and ideology. In short, epistemology, if it is to be realistic (not just realist), must be not only structural but also psychological, sociological, and historical.

Let us not forget the psychology of knowledge, a discipline which, like every other chapter of psychology, deals with one aspect of the brain activity of the higher vertebrates. Nor should we forget that there is an embryo evolutionary epistemology. Both are closely related to neurobiology and evolutionary biology. And the latter is about biopopulations, not about processes in themselves independent of the things that evolve. To be sure Popper has written about evolutionary epistemology and has asserted that his is one (Popper, 1972). But this claim contradicts his other claim, that epistemology must ignore the knowing subject. Knowledge in itself, as conceived by Popper – i.e. as an "inmate of World 3" – *does not evolve*: it is a product dwelling in the realm of Platonic ideas. Only an epistemology concerned with knowing subjects – and not only human ones – can be evolutionary.

8.8. CONCLUSION

Popper defends epistemological realism and criticizes philosophers who have held that epistemology is concerned with the beliefs of the knowing subject, instead of studying what he investigates (finds out, constructs, criticizes, etc.). He also insists that genuine knowledge is objective. In his eagerness to defend epistemological realism and the objectivity of science, Popper has also embraced metaphysical realism. In fact he has proposed two false theses.

The first thesis is that objective knowledge (and also the "contents" of works of art) constitutes a distinct and autonomous system: World 3 or the objective mind. The second thesis, or rather prescription, is that epistemology should study this "world" instead of the cognitive activities of live animals. The former thesis matches with psychophysical dualism and, by the same token, is inconsistent with biopsychology, or the study of the brain processes consisting in perceiving, imagining, thinking, remembering, and the like. And the second thesis is inconsistent with evolutionary epistemology – accepted by Popper himself – and at variance with the idea that most philosophers have about the task of epistemologists. Therefore it is unlikely to become popular.

On the other hand the many-worlds hypothesis, and in particular the thesis of the autonomy of World 3, is gaining in popularity. One reason is that it was rather popular even before being drummed up by Popper. Indeed most of us are ready to distinguish a brain from a brain process, and the latter from its "products" (e.g. constructs) — which is all right. And most of us have a tendency to separate whatever we distinguish. (We may call this fallacy the 'ontological rule of detachment'.) This is what we do when we reify — and Popper's World 3 is nothing but an instance of reification. In short the doctrine has popular appeal — and should therefore be suspect to philosophers.

Besides, Popper's trinity arrives at a time when modal logicians formalize and apply the old view that there are many, perhaps infinitely many, possible worlds. (Some of them hold that all these imaginary "worlds" are just as real as — the real world. But if pressed many of them acknowledge that such "worlds" are sets of formulas, hence even less real than the Queen and her court in *Alice in Wonderland*, which was no less than a dream "world".) This fantastic, escapist and sterile metaphysics, which encourages unbridled speculation about physically impossible worlds instead of studying reality, may have facilitated the diffusion of Popper's trinitarian metaphysics. After all, the thesis that there are just three worlds comes as a refreshing breeze of rationality after the doctrine — worthy of J.L. Borges — that there are infinitely many worlds.

In short, the three worlds doctrine is false: there is but one immensely varied and forever changing world. The thesis of the autonomous and perennial world of culture is not only false. It is also noxious, for it fosters the illusion of immortality — for the works of intellectuals and artists — even after the nuclear Armageddon. The products of cultural activity are not perennial: they are being transformed or even destroyed all the time. We have at most the possibility of cultivating the arts or the humanities, the sciences or the technologies, during our own lifetime.

And we have the duty to pass this possibility on to our children by
doing something about the only world we have got.

PART FIVE

CONCEPT

CHAPTER 9

THE STATUS OF CONCEPTS

As is well known, conceptual objects have been the undoing of traditional empiricism as well as of vulgar materialism, for they are neither distillates of ordinary experiences nor material objects or properties thereof. To be sure, the empiricist may claim that there are no conceptual objects aside from mental events. But he cannot explain how different minds can grasp the same conceptual objects and why psychology is incapable of accounting for the logical, mathematical and semantical properties of constructs. And the vulgar materialist (nominalist) will likewise discard conceptual objects and speak instead of linguistic objects − e.g. of terms instead of concepts and of sentences instead of propositions. But he is unable to explain the conceptual invariants of linguistic transformations (e.g. translations) as well as the fact that linguistics presupposes logic and semantics rather than the other way round. Therefore we cannot accept either the empiricist or the nominalist reduction (elimination) of conceptual objects any more than we can admit the idealist claim that they are ideal beings with an autonomous existence. We must look for an alternative consistent with both ontological naturalism and semantical realism.

The goals of this chapter are (a) to stress the difference between conceptual objects and material ones, (b) to characterize conceptual objects, (c) to define an existence predicate independent of the existential quantifier, and (d) to sketch a philosophy of mathematics that unites materalism with conceptualism and fictionism. All this will be done schematically and with the help of modest formal tools.

9.1. PHYSICAL AND CONCEPTUAL OBJECTS

We shall assume that there are things, or physical objects, and constructs, or conceptual objects. We shall also postulate that no thing is a construct and no construct is a thing. In other words, we shall divide every nonempty set of objects into two disjoint subsets: a set of physical objects and another of conceptual objects (either of which may be empty). Moreover we shall not assign conceptual objects the same kind of existence that physical objects possess. In fact we shall hold that the existence of conceptual objects consists in the possibility of their being thought up by some living rational being. But this will come later, after having clarified the difference between physical and conceptual objects.

The thing/construct dichotomy implies that constructs have properties essentially different from those possessed by things. In other words physical objects, whether natural or artificial, living or inanimate, share certain properties that no construct possesses. Among them we recall those of being able to change (Chapter 2), of possessing energy, of associating with others to form concrete systems possessing emergent properties, and of being localizable (though not necessarily at points in space).

The sciences that study physical objects, or things, are the factual sciences and ontology. These disciplines try to find the laws of such objects, in particular their laws of change, such as equations of motion, field equations, chemical reaction equations, social mobility matrices, and historical trends. These regularities are formulated as nomological statements. And the latter may be conceived of as restrictions upon state variables, or predicates representing (potential or actual) properties possessed by the things concerned — whether atoms or fields, cells or societies.

In other words, the law statements that factual scientists are after tell us what the really possible states of things are as well as what the really possible changes of state (i.e. events or processes) of concrete objects are. On the other hand the factual sciences do not attempt to attribute logical, mathematical, or semantical properties or laws to concrete objects. Only concepts and propositions have a sense; only propositions and (interpreted) theories can be true or false to some extent; and only theories can be logically consistent. Strictly speaking, being is meaningless. It makes no literal sense to speak of 'the meaning of life' or 'the meaning of history' unless a nonsemantic acceptation of 'meaning' is involved in these expressions. (Thus Max Weber spoke of the 'meaning of social relations' to indicate that social intercourse is purposive.)

The notions of state and change of state are central to science and ontology but alien to formal science (logic, mathematics, and semantics). For example, it makes no sense to ask about the state (mechanical, chemical, physiological, economic, political, etc.) of a number, and even less about its changes of state. Nor does it make sense to speak of the equation of motion of a Boolean algebra or of the transmutation schema of a topological space. Not that these conceptual objects are immutable: the categories of change do not apply to them, hence nor do the categories of nonchange. (Likewise it is not that culture has a zero temperature: it has no temperature at all.)

In other words constructs are in no state whatever and therefore cannot change their state. (I.e. the state space of any construct is empty.) Therefore the laws of conceptual objects do not involve the notions of state or of change of state. Sets are neither at rest nor in motion, mathematical functions are neither fertile nor sterile, algebraic structures are neither hungry nor sated, theories are neither exploiters nor exploited.

Constructs have peculiar properties that no physical object has, namely predicates. (Physical objects have or possess properties, and some properties can be variously represented by predicates. In other words, the concept of property representation can be

elucidated as a partial function from the set of thing properties
to the power set of the set of attributes. Needless to say, idealists
do not need the property/attribute distinction.) Predicates and the
propositions formed with them have meanings, but meaning has
no (physical) being. Consequently the laws satisfied by constructs
are very different from physical, chemical, biological, or social
laws. For one thing the laws of conceptual objects make no reference to physical objects and involve no state variables. Examples:
"If p, then p or q"; "If set A is included in set B, then the intersection of A and B is not empty provided B is not empty"; "If m
and n are any real numbers, then $(m+n) \cdot (m-n) = m^2 - n^2$";
"If f: $\mathbb{R} \to \mathbb{R}$ is such that $f(x) = x^2$ for any $x \in R$, then $df/dx =$
$2x$". There is nothing material about these laws; none of the ob-
jects occurring in them can be said to be in any state, let alone to
undergo a change of state.

A class of constructs of particular interest to the philosopher is
that of existence postulates. In mathematics an existence postulate
is not a refutable conjecture such as the physical hypothesis of the
existence of quarks, or partons, or elementary particles of some
other type. In mathematics an existence postulate *stipulates* that,
in the context of some theory, there are objects with such and
such characteristics. For example, in plane Euclidean geometry
one postulates that, through a point not on a straight line, there is
exactly one parallel to the given line. This postulate creates by *fiat*
the object or objects concerned, subject only to the condition that
the principle of non-contradiction be respected. (Axiomatic definitions are therefore sometimes said to be *creative*. For example,
the axiomatic definition of an abstract group creates an arbitrary
set equipped with a binary operation and a unary operation, and
including a unit element.) It is only once the (conceptual) existence of certain mathematical objects has been admitted (by
postulate) that one can proceed to postulate or prove further
properties of them. For example, before attempting to solve a

difficult differential equation it may be convenient to make sure
that it has solutions, i.e. that the latter exist (even though we may
not know them yet).

But what is meant by the expression 'Such and such an object
exists conceptually', or the phrase '*There is* an object with such
and such properties'? This is what we shall try to find out presently.

9.2. WHAT IS CONCEPTUAL EXISTENCE?

The traditional philosophies of logic and mathematics are Platonism (or objective idealism), nominalism (or vulgar materialism),
and empiricism (or psychologism). According to the former conceptual objects exist by themselves and may be embodied and
thought. To nominalism (or signism) conceptual objects are
nothing but signs or marks such as written symbols. And according
to psychologism conceptual objects are thoughts.

Each of these philosophies has both virtues and fatal flaws.
Platonism allows freedom but populates the universe with ghosts
and is therefore inconsistent with a naturalistic ontology. Nominalism emphasizes rightly the importance of language but is incapable
of distinguishing the meaningful signs from the meaningless ónes,
and it renders mathematics a game. And psychologism reminds us
that constructs are not found ready-made in some quarry, but it
does not allow us to speak of actual infinities, since nobody can
effectively think of each and every member of an infinite set.
Therefore none of the traditional philosophies of mathematics is
viable. We must therefore explore alternatives.

The alternative I should like to explore here may be called *conceptualist and fictionist materialism*. The distinctive theses of this
new philosophy of the conceptual are as follows.

(i) Conceptual objects are neither material nor ideal in the

Platonic fashion; nor are they psychical (neurophysiological) events or processes. Constructs have properties of their own, such as logical and semantical properties, that are neither physical nor mental. This is a *first conceptualist thesis*.

(ii) Conceptual objects exist in a peculiar manner, namely conceptually. More precisely, a conceptual object exists just in case it belongs to some context (e.g. a theory). Moreover it exists only as such. (For example, integers exist in number theory but not in the abstract theory of groups.) This is a *second conceptualist thesis*.

(iii) Conceptual existence, far from being ideal (Platonism), material (nominalism), or mental (psychologism) is *fictitious*. We pretend that there are sets, relations, functions, algebraic structures, spaces, etc. That is, when inventing (or learning or making use of) conceptual objects we assign them their peculiar mode of being: we demand, stipulate, feign that they exist. This is the *fictionist thesis*.

(iv) Conceiving of a conceptual object and assigning it conceptual existence are two sides of a single process occurring in some brain. Conceptual objects are thinkable and their ontological status is the same as that of mythical characters: they exist the same way that Minerva, Quetzalcoatl, or Donald Duck exist. They will cease to exist the day we stop thinking about them or imagining that they are thinkable — just as the gods of lost religions have ceased to exist. This does not entail that conceptual objects, be they mathematical, mythological, or of some other type, must be actually thought up by somebody: to exist conceptually it is necessary and sufficient to be think*able*. For example there are (conceptually) infinitely many integers that will never be thought of although every one of them is thinkable. Maybe the number 7,753,912,650,836,471,580,077,231,724,333,019,010,832 has never been thought before, but it existed by virtue of being thinkable. The same holds for all other conceptual objects. In short, the conceptual existence we assign logical, mathematical, mythical, and other such objects consists in the *possibility of being thought* of by living beings. This is the *materialist thesis.*

The preceding four theses, though supported by Section 9.1, constitute just a sketch of our philosophy of the conceptual. This sketch should be expanded into a full fledged theory in response to questions such as the following ones. What are the peculiarities of the languages in which theories of the conceptual are couched? What are the formal (mathematical) peculiarities of the basic types of construct, such as predicates, propositions, and theories? What are the semantic peculiarities of pure conceptual objects in contrast with those employed in the factual sciences? (E.g. what are formal meaning and formal truth?) What is the difference between mathematical and mythological objects? What are the relations between the mathematical concepts of a factual theory and the things it refers to?

Formal logic and the theory of models (or the semantics of mathematics) answer some of the preceding questions, though perhaps not fully. The others are still open problems. They will have to be solved, at least temporarily, by any respectable philosophy of mathematics.

9.3. CONSTRUCTS AS CLASSES OF THOUGHTS

If we want to do logic or mathematics we must pretend that there are constructs such as predicates. There is no harm in pretending that they exist, provided we do not mistake conceptual existence for material or real existence: what is real is the mental (brain) process of thinking of constructs or of any other immaterial objects, such as properties (rather than of propertied things) and mythological characters. In particular, the number 3 does not exist (by itself), though thinking of 3 is of course a process in a real thing, namely a brain.

We can do better than to espouse mathematical fictionism,

namely try and give a psychobiological account of constructs on the basis of our theory of mind (Chapter 5). Consider two or more different thoughts of one and the same object — say different instances of thinking of the number 3, or of a sphere. We can assume that all such instances, whether in the same brain or in different brains, differ in some respects only — i.e. that they are equivalent. (If they were not equivalent then they would not be instances of thinking of the same construct.) Moreover, in line with Chapter 5 we assume that each thought process is representable as a set of values of the state function $\mathbb{F}$ of the plastic neural supersystem of the person(s) concerned. And we further assume that the only difference between two instances of thinking of a given construct lies in the values of the free parameters occurring in the corresponding state functions. That is, we make the

POSTULATE. Let θ_a and θ_b be two thoughts (in a given animal at different times or in two different animals) representable by collections of values of the state functions $\mathbb{F}_a$ and $\mathbb{F}_b$ respectively. Then θ_a and θ_b are *equivalent* if, and only if, $\mathbb{F}_a$ differs from $\mathbb{F}_b$ only by the values of some of their free parameters. Symbol: $\theta_a \sim \theta_b$.

DEFINITION 1. Let θ_k be a thought and let $\sim$ be the equivalence relation defined by the above Postulate. Then the equivalence class of θ_k under $\sim$ is called the kth *construct*: $c_k = [\theta_k]_\sim$.

In ordinary language: *Every construct is an equivalence class of thoughts* (brain processes of a certain type).

Our construal of constructs is materialist because it is rooted in the notion of thoughts as brain processes, but it is not psychologistic because constructs are not equated with thoughts but rather with classes of (possible) brain processes. This secures the autonomy of logic, semantics, and mathematics vis à vis psychology.

Definition 1 above renders the intemporality of constructs plain: every thought process advances in time, but in defining a construct we abstract from time as well as from physiological details. So, we attain a position similar to Plato's theory of ideas – except that now there are no ideas without brains.

It might be objected, though, that constructs do change: thus the von Neumann construction of natural numbers differs from that of Peano, which is in turn different from that of Pythagoras, which in turn differs from the notions held by his predecessors. Granted. But the point is that, whereas material objects change *by themselves* even if we attempt to keep them unchanging, constructs do not: it is people who change, thinking now of a construct in one fashion, now in a different one. In other words, conceptual change is ultimately a change in someone's brain. If conceptual objects could change by themselves, we should be able to write down (and check) their equations of evolution (e.g. the equation of motion of an equation of motion). But it makes no sense to ask what the rate of change of a construct is, let alone what forces elicit such change.

9.4. QUANTIFICATION AND EXISTENCE

Let us now approach the technical problem of formalizing the concepts of conceptual existence and physical existence distinguished in the previous sections.

The ordinary language expressions 'there is' and 'there are' are ambiguous, for they designate two different concepts: the logical concept *some* and the ontological concept of *existence*. Logic takes care of the former and formalizes it as the existential quantifier $\exists$, which I prefer to call a *particularizer* or *indeterminate quantifier*, to distinguish it from the *universalizer* (or universal quantifier) as well as from the *individualizer* (or descriptor).

To be sure, all mathematical logicians, from Russell to Quine, have claimed that ∃ formalizes both the logical concept "some" and the ontological concept "exists". Unfortunately (a) they offer no arguments for the thesis of the identity of both concepts and (b) they are mistaken. One example will suffice to show the need for unfusing the two notions.

Consider the proposition $(\exists x)(Sx \ \& \ Bx)$, where '$S$' is interpreted as "is a siren" and 'B' as "is beautiful". That formula is generally read in either of the following ways:

(1) "There are beautiful sirens".

(2) "The schema '$Sx \ \& \ Bx$' is satisfied (is true) under the given interpretation for some values of x"

(3) "Some sirens are beautiful".

Although (1) and (2) are different — since while the former is a statement the latter is a metastatement — one might argue that they are equivalent. I.e. if there are beautiful sirens then "$Sx \ \& \ Bx$" is true (under the given interpretation) for some x, and conversely. For this reason we shall say that either of them constitutes the *ontological interpretation* of the given formula.

The ontological interpretation has in this case an obvious flaw: it suggests that the subject (writer, speaker) believes that there are sirens in reality. Most likely he only wants to say 'Some of the sirens *existing in Greek mythology* are beautiful'. The particularizer ∃ formalizes the prefix 'some' but not the expression 'existing in Greek mythology'. (The juxtaposition of two particularizers bearing on the same variable generates an ill formed formula.) We need then an exact concept of existence other than ∃ if we wish to formalize phrases such as *There are some beautiful sirens*. Let us proceed to introduce it.

We shall define a concept of relative or contextual existence, which is the one occurring in the propositions "Birds exist in nature but not in mathematics", and "Disjunction exists in logic but not in nature". We can do this by making

DEFINITION 2. Let A be a well formed set included in some non-empty set X, and χ_A the characteristic function of A, i.e. the function from X into $\{0, 1\}$ such that $\chi_A(x) = 1$ if, and only if, x is in A, and $\chi_A(x) = 0$ otherwise. Then

   (i) *x exists in A* $=_{df}(\chi_A(x) = 1)$;

   (ii) *x does not exist in A* $=_{df}(\chi_A(x) = 0)$.

To be sure we could have stipulated simply that x exists in A iff x belongs to A. But the membership relation is not a function and therefore does not allow us to take the next step.

We now introduce an existence predicate:

DEFINITION 3. The *relative* (or *contextual*) *existence predicate* is the proposition-valued function E_A from a set A to the set of all statements containing E_A, such that "$E_A(x)$" is true if and only if $\chi_A(x) = 1$.

Therefore we can now see that the old and vexing question whether existence is a predicate is ambiguous: the answer depends on whether we refer to $\exists$ or to E_A. While the particularizer is not a predicate (or propositional function, or statement-valued function), the concept of relative existence defined above is a genuine predicate.

9.5. OF HORSES AND CENTAURS

We are now in a position to distinguish two specific concepts of existence, namely those of conceptual existence (or existence in a conceptual context) and real existence (or existence in the world). We define them thus:

DEFINITION 4. If x is an object then

   (i) *x exists conceptually* $=_{df}$ For some set C of constructs, $E_C x$;

   (ii) *x exists really* $=_{df}$ For some set Θ of things, $E_\Theta x$.

For example the Schrödinger equation exists in the sense that it belongs to quantum mechanics. (It would make no sense whatever if it were a stray formula. In general, stray formulas designate no conceptual objects; only systemicity confers meaning.) Certainly it did not come into existence until Erwin Schrödinger invented it half a century ago; but it has existed ever since, though of course conceptually. Likewise the concept of an electron has existed for eight decades although its referent, the real electron, has presumably existed for ever.

The following examples show how to handle the concepts of conceptual and physical existence and how to combine them with the logical concept of some. In them 'M' stands for the set of characters in Greek mythology, 'G' for that of Greek history, 'c' for Chiron (the wisest of centaurs), 'b' for Bucephalus (Alexander's war horse), 'C' for "is a centaur", 'W' for "is wise", and 'H' for "is a horse".

The horse Bucephalus exists in Greek history.
Hb & $\chi_G(b) = 1$ or Hb & $E_G b$.
The centaur Chiron exists in Greek mythology.
Cc & $\chi_M(c) = 1$,   or Cc & $E_M c$.
Some of the individuals (existing) in Greek history are horses.
$(\exists x)(Hx$ & $\chi_G(x) = 1)$,   or $(\exists x)(Hx$ & $E_G x)$.
Some of the centaurs (existing) in Greek mythology are wise.
$(\exists x)(Cx$ & Wx & $\chi_M(x) = 1)$,   or $(\exists x)(Cx$ & Wx & $E_M x)$.
All of the characters of Greek history are real and none is mythical.
$(x)(\chi_G(x) = 1 \Rightarrow \chi_M(x) = 0)$,   or $(x)(E_G x \Rightarrow \neg E_M x)$.
All of the centaurs exist in Greek mythology and none of them is real.
$(x)(Cx \Rightarrow (\chi_M(x) = 1$ & $\chi_\Theta(x) = 0))$, or
$(x)(Cx \Rightarrow (E_M(x)$ & $\neg E_\Theta(x)))$.

So much for our analysis of the existence concepts designated by the ambiguous expressions 'there is' and 'there are'. The distinctions we have proposed allow one to remove such ambiguities. They are unnecessary when the context is fixed, as is the case with logic and mathematics, which deal only with conceptual objects. But they come in handy when the context is formed by both conceptual objects and physical ones, as is the case with the factual sciences and philosophy, in particular ontology and epistemology. Thus four of the most influential philosophies can be summarized as follows.

Vulgar materialism $(x)E_\Theta x$

Immaterialism $(x)\,^\daleth E_\Theta(x)$

Hylomorphism $(\exists x)(E_\Theta x \,\&\, E_C x)$

Conceptualist materialism $(\exists x)E_\Theta x \,\&\, (\exists y)E_C y$.

9.6. CONCLUDING REMARKS

We have sketched a philosophy of the conceptual, or conceptology, that is conceptualist, fictionist, and materialist. It is conceptualist because it admits the existence of conceptual objects distinct from physical objects (e.g. signs of a language), mental objects (which to a materialist are brain events), and ideal or Platonic objects (which to a materialist are nonexistent). Our conceptology is fictionist because, far from postulating the autonomous or independent existence of conceptual objects, it only postulates that such objects are fictitious, though not all of them idle or introduced only for the purpose of entertaining, moving, edifying, or scaring. And it is materialist because it assumes that such fictions are created and maintained by living beings, namely by merely being able to be thought about.

In order to be able to speak with some precision of the existence of conceptual objects we have had to give up the received belief that the so-called existential quantifier exactifies the notion of existence, which in turn would be one. We have introduced an existence predicate that may be specified to indicate either conceptual existence (or belonging to some set of constructs) or physical existence (or belonging to some set of concrete objects).

While conceptual existence (of e.g. a function) is either postulated or proved, physical existence (of e.g. a new particle or a new social system) is conjectured, and it is understood that such a hypothesis must be put to empirical tests. In the former case we make believe that something exists (belongs to some body of constructs) and in the latter case we assume and then verify (or refute) that some thing is part of the physical world.

Such differences notwithstanding, existence statements, whether in formal science or in factual science, are supposed to be responsible. That is, in these fields one does not waste time inventing idle constructs, i.e. concepts or propositions that discharge no function.

On the other hand, the differences between the conceptual and the material are profound with regard to the type of existence as well as the existence conditions and criteria. It is not the same thing to hold that a given differential equation has solutions of the wave type and to claim that such solutions represent real waves, i.e. chunks of matter. While the former claim can be checked with paper and pencil (and some brains), the latter requires, in addition, inventing and constructing certain special detectors and performing certain measurements with their help.

CHAPTER 10

LOGIC, SEMANTICS, AND ONTOLOGY

Philosophers have argued untiringly, over centuries, about the ties of logic with ontology. While some have followed Parmenides in identifying the two, others — particularly since Abelard — have asserted the ontological neutrality of logic. There are materialists in both camps.

Unfortunately it has seldom been made clear exactly what is meant by the 'ontological commitment' (or else 'neutrality') of logic. Is it mere reference to extralogical objects, or the presupposition of definite ontological theses, or the ontological interpretation of logical formulas, as in the case of "There are F's"?

Worse, an adequate tool for investigating this problem, namely a full-fledged semantical theory, has not been available. Indeed the only generally accepted semantical theory is model theory, or the semantics of logic and mathematics. And this theory is incompetent to handle our problem because it is solely concerned with the relations between an abstract theory (such as general group theory) and its models (e.g. the system of integers), as well as with the relations among the latter. In particular, model theory contains no theory of factual (external) reference, no theory of sense, and no theory of factual truth.

Much the same holds for semantics, though with a remarkable difference. If semantics presupposes logic, and the latter is ontologically committed, so must semantics be. But of course semantics could be tied to ontology even if logic were neutral. Therefore we need an independent investigation of the ontological commitment, if any, of semantics.

The purpose of this chapter is to investigate the relations of

logic and semantics to ontology with the help of a theory of meaning formulated elsewhere (Bunge, 1974a, 1974b). We shall summarize this theory here, so this chapter will be self-contained.

We assign meanings to constructs, in particular predicates and
propositions, and distinguish two meaning components: sense and
reference. The *sense* of a construct p in a context $\mathbb{C}$ is the totality
of logical relatives of p in $\mathbb{C}$. If p happens to belong to a theoretical
context then the sense of p in $\mathbb{C}$ is the collection of statements
within the theory that either entail p or are entailed by p. And the
reference class of a construct p in $\mathbb{C}$ is the totality of individuals
"mentioned" (truthfully or not) by p. Finally the *meaning* of p in
$\mathbb{C}$ is the ordered pair constituted by the sense of p and the reference of p. We shall apply these ideas to find out the meaning of
the typical constructs of logic and semantics. But before doing
so we must formulate those ideas somewhat more carefully. And
even before we tackle this task we must explain what we mean by
a predicate and by a context.

10.1. PREDICATE AND CONTEXT

A predicate, or propositional function, is a statement-valued function, i.e. a function that maps objects of some kind into statements
(propositions). For example, the predicate "Heavy" maps bodies
into statements of the form "b is heavy". And the propositional
function associated with the *log* function is a function mapping
pairs of real numbers into statements of the form "$\log a = b$". (I.e.
the predicate corresponding to the numerical function log: $\mathbb{R}^+ \to \mathbb{R}$
is $L: \mathbb{R}^+ \times \mathbb{R} \to S$ such that $Lxy = (\log x = y)$, where $x \in \mathbb{R}^+$ and
$y \in R$, and $\mathbb{R}$ is the real line.)

We can express this in general terms provided we assume that
the notion of a statement (proposition) has been elucidated
before. We propose

DEFINITION 1. Let A_i, with $1 \leqslant i \leqslant n$, be sets of objects of any kinds, and let S be a set of statements. Then P is a *predicate* (or propositional function) with *domain* $A_1 \times A_2 \times \ldots \times A_n$ if, and only if, $P: A_1 \times A_2 \times \ldots \times A_n \to S$, where S is the set of all those and only those statements involving P.

This non-Fregean definition of a predicate allows one to avoid Frege's confusion between *Bedeutung* (reference?, meaning?) and truth value, a confusion that invites the conflation of reference with extension and requires assuming some theory of truth before knowing what can be true or false.

Let us now turn to the notion of a context, which occurs in our definition of sense. It is introduced by

DEFINITION 2. The triple $\mathbb{C} = \langle S, \mathbb{P}, D \rangle$ is a *context* (or *framework*) iff D is a domain of individuals, $\mathbb{P}$ a set of predicates whose domains are D^n, with $n \geqslant 1$, and S a set of statements in which only members of $\mathbb{P}$ as well as logical predicates occur.

For example the loosely structured universe of discourse of a discipline is actually a context, not just a set of individuals. Every one of the behavioral sciences has its own framework even though all of them concern the same individuals.

A slightly more structured context is one which is closed under the logical operations, so that it contains whatever could be said about a subject:

DEFINITION 3. The triple $\mathbb{C} = \langle S, \mathbb{P}, D \rangle$ is a *closed context* (or *framework*) iff it is a context and if S is closed under the logical operations.

Thus the collection of propositions in a given field of inquiry, plus their denials, pairwise disjunctions and conjunctions, as well as their generalizations and instantiations, belongs to a closed context.

Obviously a closed context is partially ordered by the relation $\vdash$ of entailment. Moreover any two statements in such a context have both an infimum (their conjunction) and a supremum (their disjunction). Whence the

LEMMA: Every closed context is a complemented lattice.

Consequently we may define an ideal as well as its dual, i.e. a filter, on any closed context. And this will lead us straight on to our concept of sense.

10.2. THE SENSE OF A TAUTOLOGY

Consider a proposition p belonging to a closed context $\mathbb{C} = \langle S, \mathbb{P}, D \rangle$. Since the latter is a lattice with respect to the entailment relation, p will induce its own ideal, called the *principal ideal* of p : $(p)_{\mathbb{C}} = \{x \in S \mid x \vdash p\}$. Similarly the *principal filter* of p in $\mathbb{C}$ is $)p(_{\mathbb{C}} = \{y \in S \mid p \vdash y\}$. The same holds, mutatis mutandis, for every predicate in $\mathbb{C}$: in this case the above are sets of predicates. In either case $(p)_{\mathbb{C}}$ is the collection of logical ancestors (or determiners) of p, whereas $)p(_{\mathbb{C}}$ is the logical offspring of p. And the union of the two constitutes the totality of logical relatives of p, which we regard as the full sense of p. Therefore we make

DEFINITION 4. Let $\mathbb{C} = \langle S, \mathbb{P}, D \rangle$ be a closed context and let p be either a predicate or a proposition belonging to $\mathbb{C}$. Then
   (i) the *purport* (or upward sense) of p in $\mathbb{C}$ equals the principal ideal of p in $\mathbb{C}$, i.e. $(p)_{\mathbb{C}}$;
   (ii) the *import* (downward sense) of p in $\mathbb{C}$ equals the principal filter of p in $\mathbb{C}$, i.e. $)p(_{\mathbb{C}}$;
   (iii) the *full sense* of p in $\mathbb{C}$ equals the union of the purport and the import of p in $\mathbb{C}$:

$$\mathscr{S}_{\mathbb{C}}(p) = (p)_{\mathbb{C}} \cup)p(_{\mathbb{C}} .$$

The full sense of a construct is then the totality of constructs of the same type (i.e. predicates or statements as the case may be) occurring in the context in question and such that they entail the given construct or are entailed by it. Consequently in order to discover the sense of a proposition (or at least part of it) we must start by placing it in some closed context — preferably but not necessarily a theory; and must then proceed to find out all the assumptions from which it hangs as well as all (or at least some) of its consequences. Likewise for predicates. Therefore only theories — rather than either experiments or philosophical vagaries — will tell us what the sense of theoretical concepts are. For example, to discover the sense of "electric charge" we resort to electrodynamics, not to experiment, let alone to the psychology of electric shocks.

Now, an arbitrary statement entails, among other propositions, all the tautologies found in the logic associated with the context in which it occurs. Likewise every predicate implies any tautological predicate. If we wish to retain only the synthetic or extralogical part of the sense of a construct, we must subtract logic from it:

DEFINITION 5. Let $\mathbb{C} = \langle S, \mathbb{P}, D \rangle$ be a closed context and L the logic associated with it. If p is a construct in $\mathbb{C}$, then the *extralogical sense* of p in $\mathbb{C}$ is

$$\mathscr{S}_{\bar{L}\mathbb{C}}(p) = \mathscr{S}_{\mathbb{C}}(p) - L,$$

i.e. whatever is in the full sense but not in the logic.

Clearly, whatever sense a construct has in addition to its logical sense is determined by all the nontautological constructs it is related to within the context concerned, and in the first place by the primitive or basic constructs (which constitute the *gist* of the construct).

We are now equipped to face the problem of determining the sense of an analytic statement. Firstly we have

THEOREM 1. *Let t be a tautology belonging to the logic L underlying a closed context $\mathbb{C} = \langle S, \mathbb{P}, D \rangle$. Then*

(i) *the purport (upward sense) of t equals the totality S of propositions in $\mathbb{C}$*:

$$\mathscr{Pu}_{\mathbb{C}}(t) = S;$$

(ii) *the import (downward sense) of t equals the collection of all tautologies in L, i.e. L itself:*

$$\mathscr{Imp}_{\mathbb{C}}(t) = L;$$

(iii) *the full sense of t in $\mathbb{C}$ is*

$$\mathscr{S}_{\mathbb{C}}(t) = S \cup L;$$

(iv) *the extralogical sense of t in $\mathbb{C}$ equals the nonlogical part of S:*

$$\mathscr{S}_{\bar{L}\mathbb{C}}(t) = S - L.$$

Proof: The first part follows from the fact that t is entailed by any proposition p in S; i.e. every $p \in S$ is in the purport of t. The second part follows from the interdeducibility of all the tautologies in a given L. The third follows with the assistance of Definition 4 (iii) and the fourth with the help of Definition 5.

COROLLARY 1. Let $\mathbb{C}$ be a purely logical context, i.e. $S = L$. Then the extralogical sense of a tautology $t \in S$ is nil:

$$\mathscr{S}_{\bar{L}\mathbb{C}}(t) = \emptyset.$$

In other words, within logic tautologies do not "say" anything that is not strictly logical. It is only when meshing in with a body of substantive knowledge that logic does "say" something – in fact too much.

The dual of Theorem 1 concerns contradictions:

THEOREM 2. *Let* $\mathbb{C} = \langle S, \mathbb{P}, D \rangle$ *be a closed context with underlying logic L and let t be a tautology of L. Then*

(i) *the purport of* $\neg t$ *in* $\mathbb{C}$ *is nil:*

$$\mathscr{P}\!u\!_{\mathbb{C}}(\neg t) = \emptyset \, ;$$

(ii) *the import of* $\neg t$ *equals the totality S of propositions in* $\mathbb{C}$:

$$\mathscr{I}\!m\!p_{\mathbb{C}}(\neg t) = S \, ;$$

(iii) *the full sense of* $\neg t$ *in* $\mathbb{C}$ *is:*

$$\mathscr{S}_{\mathbb{C}}(\neg t) = S \, ;$$

(iv) *the extralogical sense of* $\neg t$ *in* $\mathbb{C}$ *is the set of nonlogical statements in* $\mathbb{C}$:

$$\mathscr{S}_{\overline{L}\mathbb{C}}(\neg t) = S - L \, .$$

Proof. The first part follows from the tacit assumption that S is a consistent set of statements; if on the other hand S were to contain contradictions, these would be in the purport of $\neg t$ because all contradictions are interdeducible. The second part follows from the fact that a contradiction entails anything. The third and fourth follow with the help of Definitions 4 (iii) and 5 respectively.

COROLLARY 2. Let $\mathbb{C}$ be a purely logical context, i.e. $S = L$. Then the extralogical sense of a contradiction $\neg t$, where $t \in L$, is nil:

$$\mathscr{S}_{\overline{L}\mathbb{C}}(\neg t) = \emptyset \, .$$

In sum, tautologies and their negates have a definite sense which depends upon the context in which they occur. If the context is purely logical (e.g. a logical calculus) so is the sense of the tautology. A tautology "says" something provided it is embedded in a body of substantive knowledge, for then it depends on every proposition included in that body. But what it "says" does not go beyond what that body of knowledge states. And isolated from every body of substantive knowledge, i.e. in itself, logic "says" nothing extralogical — as could not be otherwise.

In every context $\mathbb{C} = \langle S, \mathbb{P}, D \rangle$ we have then a minimal sense, namely L, and a maximal sense, viz. $L \cup S$. The sense of a construct that is neither tautologous nor contradictory lies between these extremes. More precisely, we have

COROLLARY 3. Let p be a construct in a closed context $\mathbb{C} = \langle S, \mathbb{P}, D \rangle$ with underlying logic L. Then the full sense of p is comprised between the intralogical sense and the full sense of a tautology in the same context, i.e.

$$L \subseteq \mathscr{S}_{\mathbb{C}}(p) \subseteq S \cup L.$$

For example, the sense of "probability", within pure mathematics, includes all the items in the calculus of probability that involve the probability function. But the sense, hence the concept itself, swells if placed in the context of applied probability theory. Another example: contrary to what an extensionalist must suppose, "round square" is meaningful. If it were nonsensical just because its extension is nil, we would be unable to assert the nonexistence of round squares. The statement "There are no plane figures both round and square" is provable in plane geometry, hence it has a nonempty purport; and it also has a nonvoid import: it "says" that points, triangles, etc. are not both round and square.

So much for the sense of tautologies and contradictions. We turn now to the second component of meaning.

10.3. THE REFERENCE OF A TAUTOLOGY

The referents of "Terra lies between the Sun and Jupiter" are the three bodies in question. In general, if a predicate P "applies" (truthfully or not) to objects $a_1, a_2, \ldots, a_n$, then these are its referents. That is,

$$\mathscr{R}(Pa_1 a_2 \ldots a_n) = \{a_1, a_2, \ldots, a_n\}.$$

Clearly, the denial of a statement does not change its reference class even though it alters its truth value. And if a second statement combines with the first either disjunctively or conjunctively, it contributes its own referents. That is, the reference function $\mathscr{R}$ is insensitive to the propositional connectives:

$$\mathscr{R}(\neg p) = \mathscr{R}(p), \quad \mathscr{R}(p \vee q) = \mathscr{R}(p \& q) = \mathscr{R}(p) \cup \mathscr{R}(q)$$

for any propositions p and q. Similarly for predicates. Not surprisingly, $\mathscr{R}$ is also insensitive to the precise kind of quantifier.

We sum up the preceding intuitive remarks in two definitions, one for the reference of predicates, the other for the reference of statements:

DEFINITION 6. The reference class of a predicate is the set of its arguments. More precisely, let $\mathbb{P}$ be a family of n-ary predicates with domain $A_1 \times A_2 \times \ldots \times A_n$. The function

$$\mathscr{R}_p : \mathbb{P} \to \mathscr{P}(\cup_{1 \leqslant i \leqslant n} A_i)$$

from predicates to the set of subsets of the union of the cartesian factors of the domains of the former, is called the *predicate reference function* iff it is defined for every P in $\mathbb{P}$, and its values are

$$\mathscr{R}_p(P) = \cup_{1 \leqslant i \leqslant n} A_i.$$

An almost immediate consequence of this definition is

THEOREM 3. *The propositional connectives concern statements.*

Proof: Apply Definition 6 to negation, then to disjunction, taking into account that the functional construals of these logical predicates are

$$\neg : S \to S, \quad \text{with} \quad \neg(s) = \neg s \quad \text{for any} \quad s \in S;$$
$$\wedge : S \times S \to S, \quad \text{with} \quad \wedge(s, t) = s \wedge t \quad \text{for any} \quad s, t \in S.$$

A further consequence is

COROLLARY 4. The reference class of a tautological predicate equals the union of the reference classes of the component predicates. In particular

(i) $\mathscr{R}_p(P \vee \neg P) = \mathscr{R}_p(P)$;

(ii) $\mathscr{R}_p((P)(P \vee \neg P)) = \cup_{P \in \mathbb{P}} \mathscr{R}_p(P)$.

As for the reference class of a statement, it will be computed with the help of

DEFINITION 7. Let $\mathbb{P}$ be a family of n-ary predicates with domain $A_1 \times A_2 \times \ldots \times A_n$ and let S be the totality of statements formed with those predicates. The function

$$\mathscr{R}_s : S \to \mathscr{P}\,(\cup_{1 \leqslant i \leqslant n} A_i)$$

is called the *statement reference function* iff it is defined for every s in S and satisfies the following conditions:

(i) The referents of an atomic statement are the arguments of the predicate concerned. More precisely, for every atomic formula $P a_1 a_2 \ldots a_n$ in S,

$$\mathscr{R}_s(P a_1 a_2 \ldots a_n) = \{a_1, a_2, \ldots, a_n\}.$$

(ii) The reference class of an arbitrary propositional compound equals the union of the reference classes of its components. More precisely, if $s_1, s_2, \ldots, s_m$ are statements in S and if ω is an m-ary propositional operation, then

$$\mathscr{R}_s(\omega(s_1, s_2, \ldots, s_m)) = \cup_{1 \leqslant j \leqslant m} \mathscr{R}_s(S_j).$$

(iii) The reference class of a quantified formula equals the reference class of the predicate occurring in the formula. More explicitly, if P is an n-ary predicate in $\mathbb{P}$, and the Q_i (for $1 \leqslant i \leqslant n$) are arbitrary quantifiers,

$$\mathscr{R}_s((Q_1 x_1)(Q_2 x_2) \ldots (Q_n x_n) P x_1 x_2 \ldots x_n) = \mathscr{R}_p(P).$$

The following consequence is immediate:

COROLLARY 5. The reference class of a tautological statement equals the union of the reference classes of the predicates involved. In particular

(i) $\qquad \mathscr{R}_s((x)(Px \vee \neg Px)) = \mathscr{R}_p(P)$

(ii) $\qquad \mathscr{R}_s((P)(x)(Px \vee \neg Px)) = \cup_{P \in \mathbb{P}} \mathscr{R}_p(P).$

For example, 'The arrow moves or does not move" refers to a certain arrow. Its universal generalization "Everything either moves or it does not" refers to any object capable of moving, i.e. to all physical objects. And its higher order generalization "Whatever the property, everything either has it or it does not" refers to the whole set Ω of objects, whether physical or mental. We call such tautologies *universal*. The second part of Corollary 5 can then be rewritten as

COROLLARY 6. Any universal analytic statement $t \in L$ in a logical theory L refers to all objects:

$$\text{If} \quad t \in L \quad \text{then} \quad \mathscr{R}(t) = \Omega.$$

Because of Definition 7(ii), the same holds for contradictions. On the other hand the extension of a contradiction is nil. This is a warning against the temptation to identify reference with extension.

So much for the second component of meaning. We are now in a position to find out the meaning of a logical formula.

10.4. THE MEANING OF A TAUTOLOGY

As we have seen, the sense function $\mathscr{S}$ maps constructs into sets of constructs, i.e. $\mathscr{S}: C \to \mathscr{P}(C)$, whereas the function $\mathscr{R}$ maps constructs into sets of objects of any kind, i.e. $\mathscr{R}: C \to \mathscr{P}(\Omega)$, where $\mathscr{P}$ is the power set function. Since the maps $\mathscr{S}$ and $\mathscr{R}$ have been defined, the function

$$\mathscr{M}: C \to \mathscr{P}(C) \times \mathscr{P}(\Omega)$$

remains uniquely determined for each type of construct. We call it the *meaning function*. In other words, we adopt

DEFINITION 8. Let $\mathbb{C} = \langle S, \mathbb{P}, D \rangle$ be a closed context and p a predicate or a proposition in $\mathbb{C}$. *Then the meaning of p in $\mathbb{C}$ is the sense of p together with the reference of p*:

$$\mathscr{M}_{\mathbb{C}}(p) = \langle \mathscr{S}_{\mathbb{C}}(p), \mathscr{R}_{\mathbb{C}}(p) \rangle .$$

Putting together Theorem 1 and Corollary 6 we get

COROLLARY 7. The meaning of a universal tautology $t \in L$ in an arbitrary context $\mathbb{C} = \langle S, \mathbb{P}, D \rangle$ is S plus logic together with the set D of all objects in $\mathbb{C}$:

$$\mathscr{M}_{\mathbb{C}}(t) = \langle S \cup L, D \rangle .$$

By Corollary 1, analytic statements, when detached from all bodies of substantive knowledge, are devoid of extralogical sense, ergo

COROLLARY 8. Within logic, universal tautologies say nothing about everything:

$$\text{If} \quad t \in L \quad \text{and} \quad S = L, \quad \mathscr{M}_{\bar{L}\mathbb{C}}(t) = \langle \emptyset, D \rangle .$$

The dual of Corollary 8 is

COROLLARY 9. Universal contradictions say anything about everything:

$$\text{If } t \in L \text{ and } \neg t \in \mathbb{C}, \text{ then } \mathscr{M}_{\mathbb{C}}(\neg t) = \langle S, D \rangle.$$

In all three corollaries, the domain D blows up to the entire set Ω for maximally general analytic statements.

Analytic propositions, in short, *are* meaningful: they have the smallest possible sense and the largest possible reference.

From the fact the universal tautologies "apply" to (are true of) anything, or "hold in every possible world", it has been concluded that logic is a kind of ontology (see e.g. Scholz, 1941; Hasenjaeger, 1966). This is an unwarranted conclusion for, although a universal tautology *refers* to anything, it *describes* nothing but logical objects such as "or" and "entails". This is what Corollaries 7 and 8 state. And it is perhaps what is meant when claiming the opposite view, namely that tautologies are meaningless. Tautologies, if our semantics is adequate, do have a sense but by themselves they *state* nothing about the world even if they refer to it.

The foregoing analysis does not disprove the thesis that logic can be assigned ontological *interpretations*. No doubt, it is admissible to interpret the individual variables as ranging over entities and the extralogical predicates as ranging over properties of entities. Upon proceeding in this fashion one obtains, for instance, the following interpretation of the excluded middle principle: "Every entity either has a given property or fails to have it". Yet this ontological interpretation does not yield an ontological *thesis*, e.g. a metaphysical axiom, but rather a retreat from every ontological commitment. Ontology is supposed to make definite and positive statements about the structure of the world instead of evading the issue as logic does. In short, the ontological interpretations of logic are harmless: far from constituting a sort of minimal ontology they are a reminder that logic is context-free, i.e. invariant under changes in contexts or fields of inquiry. Moreover, if analytic statements did say anything definite about reality they would be at the same time synthetic. Finally, even if an ontological inter-pretation of logic were to pass for ontology, it still would be some-thing other than pure logic.

However, none of the preceding arguments disposes of an even more radical thesis, namely that logic *presupposes* an ontology. But this claim can be disposed of in three sentences. Firstly, the only place where an ontology might creep into logic is by way of the notion of domain of a predicate (Section 10.1), which set must not be empty for the predicate to exist. However, and secondly, the individuals forming that set need not be specified: they could be physical, conceptual, or even totally nondescript, as they are in the case of, say, an unspecified unary predicate $P: A \to S$. Finally, the mere requirement that such domains, e.g. A, be nonempty, is not an ontological assumption but a condition for P to be *called* a predicate: indeed P won't be such unless it attributes or assigns a property to something. In sum, logic pre-supposes no ontological thesis. It is rather the other way around: any cogent ontology presupposes some logic.

Logic, in sum, is ontologically noncommittal. This is why it can be made to reign over all contexts. What about semantics? This is another story.

10.5. SEMANTICS AND ONTOLOGY

If we now apply our theory of reference to some of the typical concepts of semantics, it turns out that while some of them refer to constructs others refer to objects of any kind. For example, the reference relation maps constructs into sets of objects (subsets of

Ω), whence the corresponding predicate is the propositional function

REF: $C \times \mathscr{P}(\Omega) \to S$ such that REF$(c, A) = \ulcorner \mathscr{R}(c) = A \urcorner$ for c in C and A in $\mathscr{P}(\Omega)$. Hence $\mathscr{R}(\mathscr{R}) = C \cup \mathscr{P}(\Omega) = \mathscr{P}(\Omega)$. Likewise for the extension function.

We might conclude that semantics as a whole does refer to objects of any kind. But it does so in the same noncommittal way that logic "applies" to anything, that is, without describing or representing (let alone explaining and predicting) the behavior of anything but its specific concepts. Hence, as far as reference is concerned, semantics is just as noncommittal as logic.

It would seem then that Tarski (1944; p. 363) was right in asserting that ontology, in case it exists, "has hardly any connections with semantics". This is obviously true of the semantics Tarski had in mind, namely the semantics of mathematics, or model theory. Indeed, mathematics is the study of conceptual structures, such as lattices, number systems, manifolds, and categories, and model theory focuses on the most abstract of all structures. None of these is an entity or real thing, i.e. the object of ontology. But Tarski's dictum does not apply to the semantics of factual (empirical) science — which anyway did not exist at the time Tarski made that statement. In fact this other semantical theory meets metaphysics at least at two points: reference and truth. Let us glance at these interfaces.

Clearly, a general theory of reference needs no ontological background. But the *applications* of any such theory to the semantic analysis of factual predicates and propositions do call for certain definite assumptions as to what counts as a referent. Consider the statement "Car b stopped at point p at time t". Do b, p, and t qualify as referents? They do according to the naive ontology of the man in the street, to whom space and time are as real as cars. But alternative ontologies will come up with different identifications of the referents. In a process metaphysics there will be a single referent, namely the event of the car stopping. And in a systems metaphysics there will be two referents, namely the car and the thing located at p and t.

Furthermore the very distinction between reference in general, and factual reference (or reference to factual items) in particular, presupposes that there are facts and that they constitute a proper subset of all possible objects — an ontological assumption. That the distinction is important in the semantics of science can be shown by a simple example, such as that of mass. The concept of mass in classical physics can be construed as a function that associates every ordered triple ⟨body, time instant, mass unit⟩ with a positive real number. According to our theory of reference, the total reference class of "mass" is then the union of the set of bodies, the set of instants, and the set of mass units. Since mass units are not factual items, they do not occur in the factual reference class of the mass concept.

Ontology raises its head also in the matter of truth. Certainly the coherence theories of truth need no ontological assumption. (Incidentally the model theoretic concept of truth as satisfaction in a model belongs in this class, since it construes truth as the fitting of one conceptual structure to another.) But the concept of (factual and partial) truth used in daily life and in the sciences is the one that ought to be elucidated by a (nonexistent) correspondence theory of truth. And obviously, if an *ens rationis* is to be adequate to a *res* (thing or fact), then we must begin by assuming that *there are* such *rei* or extraconceptual objects. In other words, the view that truth consists in an *adaequatio intellectus ad rem* requires the existence of *rei*. This is surely a modest ontological assumption, yet one that non-realists, such as subjective idealists and logical positivists, do not accept. Nor do they need it: the former because the coherence theory suffices for them, the latter because they have no use for any theory of truth except perhaps the trivial theory of truth by convention.

Note finally that our last argument is independent from Quine's thesis that the mere use of the existential quantifier commits us to ontology (cf. Quine, 1969). Firstly, we were discussing semantics, not logic. Secondly, we were concerned with factual truth, not with quantification theory. Thirdly, "$(\exists x)Px$" may be taken to assert the existence of P's — but these objects may or may not be physical, depending on the interpretation of P. As it stands, the proposition is a neutral existence assertion and moreover one that could be made just for the sake of the argument. And it is just as well that a mere unqualified existential quantifier carries no ontological load if logic is to be valid regardless of the interpretation of the extralogical predicates. We do commit ourselves one way or another just if we transform the preceding existence statement into either "$(\exists x)(Px$ & x is a physical object)" or "$(\exists x)(Px$ & x is a construct)". (More in Chapter 9.) Thus the physical scientist who undertakes to investigate any sector of physical reality presupposes unwittingly that there are physical objects. So does the ontologist.

10.6. CONCLUSIONS

We have employed our semantical theory to investigate the matter of the ontological commitment of logic and semantics, and have found the following results:

(i) The logical predicates, such as "or" and "entails", are about predicates and statements, not about any other objects. Moreover this is what a logical theory is supposed to characterize, namely logical objects, and only such.

(ii) Tautologies refer, now to conceptual objects, now to physical ones, now to all objects. But they describe or characterize none. Hence logic is not *"une Physique de l'objet quelconque"* (Gonseth, 1938, p. 20). Logic is no more and no less than the theory of logical form, in particular of the form of deductive arguments.

(iii) The general theory of reference we propose does not specify the nature of the referents of a construct, hence it is not ontologically committed. By contrast, any application of the theory is ontologically committed. This commitment is made the moment the predicate under scrutiny is analyzed as a function mapping n-tuples of objects into statements. Such an analysis requires the identification of such objects, which identification is based on some hypothesis or other concerning the furniture of the world. But the identification itself is a task for the special sciences, not for semantics.

(iv) The correspondence theory of truth is committed to the thesis that there is an external world, i.e. that there are entities that a factually true statement fits. Hence the chapter of semantics dealing with the notion of *vérité de fait* is not ontologically neutral. And it does not overlap with model theory.

In a nutshell: whereas logic is ontologically neutral, semantics is partially committed to some ontology or other. Of course this result is critically dependent upon our semantical theory. Hence anyone wishing to dispute it should avail himself of an alternative semantical tool. Otherwise he won't be able to say anything with precision nor, a fortiori, to prove anything about the relations between logic and semantics, on the one hand, and the world on the other.

APPENDIX

NEW DIALOGUES BETWEEN HYLAS
AND PHILONOUS

FOREWORD

The *Three Dialogues Between Hylas and Philonous* were written by George Berkeley in 1713 as a popular exposition of his no less famous *Treatise Concerning the Principles of Human Knowledge*, published three years earlier. Since then, the basic philosophical ideas of the Bishop of Cloyne do not seem to have been satisfactorily refuted – although they have been and are still being abundantly criticized. The most usual argument against Berkeley's idealistic empiricism is still the stick employed by Molière to convince Pyrrhonists. Of course, the argument from practice is historically important and psychologically effective – but, after all, it is in keeping with Berkeley's ideas, its succes being thus at the same time a paradoxical triumph of his own empiricism.

It is probable that one reason that no conclusive logical arguments against Berkeleyanism seem to have been given is that most of the Bishop's opponents, when thinking of refuting him, tacitly accept that experience alone – for instance, experience with sticks – is able to verify or falsify a proposition. For, as long as it is believed that only facts can compete with facts; that reason is only able to reflect or at most to combine sense data, being incapable of creation; and as long as one clings to the belief that there cannot be rational proofs of empirical facts, it is only natural that Berkeley's chain of thought should remain substantially untouched. The Bishop knew this only too well, and that is why he based his system on the negation of abstract thought, on the thesis that abstract ideas are nothing but vices of language.

The present dialogues are an endeavor to refute Berkeley's

philosophy from a new standpoint. They are offered in homage to his astuteness, on the occasion of the second centenary of his death in 1753.

THE FIRST DIALOGUE

PHILONOUS. Good morrow, Hylas: I did not expect to find you alive. We have not seen each other for centuries.

HYLAS. To be exact, for two hundred and forty years. By the bye: How do you explain it that, notwithstanding, we have continued existing?

PHILONOUS. You know that the age of philosophies must be measured in centuries, not in years.

HYLAS. I did not mean that, but the following: We have not perceived each other during more than two centuries; nevertheless, each of us is certain that the other has been alive during that lapse of time.

PHILONOUS. Now I see what you mean. But your irony is out of place. The distinctive axiom of my philosophy is *To be is to perceive or to be perceived*. I did not perceive you, nor did you perceive me during that time; but you have perceived other things, and so have I — therefore, we have both existed.

HYLAS. Allow me to dwell on this. I did not see you, nor did you perceive me; so that you can be certain that *you* have existed, but you could not be sure that I was alive, say, one hundred years ago.

PHILONOUS. Why not? I am verifying it now. If I had not seen you alive, I could not maintain it with certainty. But I am seeing you now, and since I absolutely rely on my senses, I am sure of it.

HYLAS. Yes, you *verify* that now. But it has not been enough for you to see me again in order to *know* that I was alive one century ago: this knowledge was not contained within your perceptions, because senses do not make inferences. It is a product of reasoning.

PHILONOUS. I grant it. But, instead of speaking about rational inferences, I would prefer to speak about sequences of images. I do not find it difficult to imagine that you have existed one hundred years ago – or, for that matter, to imagine an hippogriph.

HYLAS. Of course you can imagine it, but you cannot *demonstrate* it unless you are able to ascend from images to concepts, because no sequence of images will ever make a proof. This is just why we form concepts and perform inferences: in order to know and to prove whenever the evidences of the senses are not enough to know and to prove – and they never are.

PHILONOUS. Could you prove, perhaps, the fact that you were alive one century ago, without resorting to evidences of an historical sort?

HYLAS. Certainly, though not by purely logical means. In order to turn it into the result of an inference, I need one more premise besides *Hylas existed in 1753 and exists now, in 1953* – the latter being true according to your own criterion of truth, since it has been warranted by the senses.

PHILONOUS. What is that new premise you need?

HYLAS. A law of nature known by induction from numberless singular cases, namely, that the life cycle of every individual is uninterrupted. In this way, the logical demonstration that I was alive in 1853 reduces itself to a simple syllogism.

PHILONOUS. I must avow that this assumption, which you wrongly call a law of nature, was underneath my inference. Of course, I will not call it a law of nature, but a rule ordained by the eternal Mind.

HYLAS. Be that as it may, the net result of our conversation is that reason is not only able to reflect sensible things, but is also capable of proving, or at least suggesting, the existence of empirical facts otherwise unknown by immediate perception. Such is the case with the fact expressed in the sentence *Hylas was alive one century ago* — which I would call a rational truth of fact.

PHILONOUS. Methinks I agree. But, in return, you must avow that it was not pure logic which gave us this result, since you had to use a so called law of nature.

HYLAS. I acknowledge it with pleasure, because I am not for a sharp dichotomy between theory and experiment. It was not pure logic which gave us that result; but it was a non-empirical procedure — profiting, of course, from previous experience and being an experience itself — wherein the laws of thought bring together facts and laws of nature. But let us return to our argument.

PHILONOUS. We had come to agree that I must accept as a truth that *Hylas existed in 1853*, notwithstanding the fact that it did not pass through my senses.

HYLAS. That was it. Now, if you admit it, you will have to avow that the famous saying that the intellect contains nothing that has previously not been in the senses, is at least partially untrue. In other words, you must acknowledge that reason is a kind of practice capable of *creating* things — things of reason, or ideal objects which, of course may refer to sensible things.

PHILONOUS. Slow my friend. A single instance does not suffice to confirm a theory.

HYLAS. But it is enough to disprove a theory — yours, for instance. Besides, if you wish I can add a host of cases showing that we cannot dispense with ideal constructs unperceivable or as yet unperceived, existing in the outer reality or in thought alone.

PHILONOUS. I will content myself with a couple of samples.

HYLAS. First, neither you nor anybody else can perceive the facts that occurred millions of years ago, as reported by geology or by paleontology — which reports, incidentally, are believed by the petroleum companies. Second, physics, astronomy and other sciences are engaged in predicting future events, events known by us if only probably, but in any case anticipating perception — and even, as in the case of scientific warfare, annihilating perception.

PHILONOUS. I own there is a great deal in what you say. But you must give me time to look about me, and recollect myself. If you do not mind, we will meet again tomorrow morning.

HYLAS. Agreed.

THE SECOND DIALOGUE

PHILONOUS. I beg your pardon, Hylas, for not meeting you sooner. All this morning I have been trying – alas! unsuccessfully – to refute your contention that the proofs of reason may be as acceptable as the evidences of the senses, and that even creations of thought may be as hard as facts.

HYLAS. I expected it.

PHILONOUS. Still, you did not convince me that we are able to form abstract ideas.

HYLAS. Nevertheless, it is an empirical fact that you cannot avoid employing abstract ideas, such as existence, being, idea, all, none, etc. – especially when trying to convey the abstraction that all abstraction is fiction.

PHILONOUS. Well, I might concede to you that we are able to form abstract ideas. But I hold that we derive them all from perceptions. To be more precise, mind can elaborate the raw material offered by the senses, but it cannot create new things, it cannot make objects that are unperceivable in principle.

HYLAS. You forget that we agreed yesterday that not every idea has a previous existence in immediate perception. Remember, this was your case with *Hylas existed in 1853*.

PHILONOUS. Yes, but that idea *might* have arisen from the senses, if only I had had a chance of seeing you one hundred years ago. Moreover, you must avow that the ideas of existence and being, which you deem to be abstract, are in any case derived by a sort of distillation of an enormous aggregate of concrete ideas of existent beings.

HYLAS. Naturally, I agree with you, and I am glad to detect a little germ of historical reasoning in you. This is just how most abstract ideas are formed: by a long distillation — though not a smooth one. But this is not the case with all abstract ideas: some are pure, though not free creations of the human mind — even when they arise in the endeavor to grasp concrete things.

PHILONOUS. Do I hear well, Hylas? Are you defending spirit?

HYLAS. I never was its enemy. It was *you* who denied the existence of abstract ideas and, in general, the possibility for mind of creating new ideas without previously passing through experience — thus reducing your famous *nous* to a poor little thing.

PHILONOUS. It will not be long before you accuse me of atheism.

HYLAS. To be sure, I could do it. Think only of the imperfections of your God, who has neither the attributes of materiality nor of abstract ideas. But let us leave theology aside: I do not wish an easy victory over you. I accuse you of being an inconsistent empiricist, because you do not understand that abstract thinking is an activity, an experience. And I accuse you of being an inconsistent idealist, because you do not understand that mental activity is able to create new objects, things that are not to be found in perception.

PHILONOUS. Pray, give me a single instance of these ideal objects not developed from perceptions and moreover, as you would say, not having a material counterpart.

HYLAS. Mathematics is full of such entities not corresponding to any objective reality but which, nevertheless, are auxiliary instruments in the labor of explaining and mastering the world. Not to mention higher mathematics, let me only recall imaginary numbers. Or, better, an even simpler object, the square root of two, or any other irrational number, which you would never obtain by measurement, by perception.

PHILONOUS. As far as I am concerned, these inventions might as well not exist. You know, I struggled against such absurdities long ago.

HYLAS. I remember: You maintained that they were not only meaningless but also harmful — by the way, there you have two nice abstract concepts, which you have employed hundreds of times: meaningless and harmful. Still, as you ought to know, events have demonstrated that irrational numbers are one of the prides of reason, and that imaginary numbers are nearly as useful and practical as real ones.

PHILONOUS. Your arguments remain very poor if they cannot pass beyond the inventions of the analysts.

HYLAS. Do not worry. You find plenty of abstract ideas outside of mathematics and logic. You always employ them: design, eternity, identity, whole... The trouble with you, Philonous, is that you move so freely on the plane of abstractions that you do not notice it and take them for granted.

PHILONOUS. Perhaps a detailed historical investigation could show that they are really not *new* ideas, but mere refinements and combinations of percepts.

HYLAS. I would not advise it: history is precisely the most effective destroyer of errors. For instance, it is history that shows how concrete singulars become abstract universals, and how the latter enable us to discover and even to make new sensible things. But let us not go astray. I tried to convince you before that there are abstract ideas not made up of perceptions and lacking a material counterpart.

PHILONOUS. Right.

HYLAS. Now I wish to remind you that, conversely, there are real things that can solely be grasped by abstraction I mean wholes and structures, of which perception and even image provide us with only partial accounts. For instance, you cannot perceive the Irish people, or democracy, or mankind. Nor can you see or smell order, law, prosperity and what not. Of course, you understand them all on the basis of percepts and with the aid of images; but they are, none the less, abstract ideas corresponding, this time, to objective wholes.

PHILONOUS. I suggest that we do not discuss my deep political convictions. Why do not we return to logic?

HYLAS. With delight. Since you feel so sure about your logic, allow me to put the following question: How do you know that perceptions are the sole ultimate and authentic source of knowledge, the sole factory of human knowledge and the sole guarantee of reality?

PHILONOUS. Proceeding along empirical lines, I have found that this bit of knowledge, and that one, and a third, they all derive from sense experience.

HYLAS. I very much doubt that you have actually proceeded that way. But, for the sake of argument, let us assume that every item of knowledge stems from a corresponding percept; that every singular cognition derives from a corresponding experience — which is in itself singular, unless you should hold that experience is capable of yielding universals.

PHILONOUS. God forbid!

HYLAS. Well, then I further ask: Whence comes this *new* knowledge, this universal judgment which you take for true, namely *The source of all knowledge is experience?*

PHILONOUS. I am not sure that I have understood you.

HYLAS. I said that, for the sake of discussion, I might grant that every singular knowledge comes from experience. But whence comes your knowledge that all knowledge comes, came or will come from sense experience? Does *this* new knowledge originate in experience too?

PHILONOUS. I avow that your argument embarrasses me. I should have dispensed with such general maxims, in the same way as I excluded abstract ideas. But, then, what would have remained of my teachings?

HYLAS. Nothing. And this is just Q.E.D. – that your entire system is false, because it relies on a *contradictio in adjecto.*

PHILONOUS. Do not abuse me in Latin, please.

HYLAS. I will explain that to you, but tomorrow, if you please. I shall be glad to meet you again at about the same time.

PHILONOUS. I will not fail to attend you.

THE THIRD DIALOGUE

HYLAS. Tell me, Philonous, what are the fruits of yesterday's meditations?

PHILONOUS. I found that you were right in suggesting that the first axiom of empiricism – *The sole source of knowledge is experience* – is an abstract idea and, to make things worse, partially untrue.

HYLAS. And self-contradictory, as I told you yesterday with a Latin jargon. In effect, the starting-point of empiricism – as of every other philosophy – is not experience but a universal judgment, so that empiricism starts denying abstraction in abstract terms, with which it destroys itself.

PHILONOUS. Thus far I am forced to avow. But I challenge you to prove the falsity of my principle *To be is to perceive or to be perceived.*

HYLAS. As far as I remember, I did it in our first dialogue. But, since you are now more used to abstractions, I will give you more refined arguments. In the first place, notice that you cannot maintain that saying on empirical grounds any longer, since we experience singulars only, never universals.

PHILONOUS. I agree.

HYLAS. My new proof runs as follows: If you admit that you are able to conceive at least *one* abstract idea, an idea not immediately derived from the senses – an idea by definition imperceptible and unimaginable – then your famous principle is done with.

PHILONOUS. I do believe now that I have always played with abstract ideas, but still I do not see your point.

HYLAS. That concession of yours implies two things. First, at least sometimes — while you are making abstractions, while you are working with concepts — you exist without being conscious of your sense impressions. And this destroys your *esse est percipere*. Second, granting the existence of abstract ideas you concede that not every thing consists in being perceived, since abstractions are unperceivable. And this does away with your *esse est percipi*.

PHILONOUS. I am forced to own it. But this new concession would demand only a slight change in my system: from now on I will say that existence is identical with *any* faculty of the mind.

HYLAS. You are wrong in supposing that you will manage to save your immaterialism after so many concessions as you have made.

PHILONOUS. Why not? So far, I have only admitted theses concerning mind.

HYLAS. That is sufficient. As soon as you admit — as you have done it — that not every thing consists in being perceived; and as soon as you accept the validity of theoretical proof, you are forced to admit at least the possibility of theoretically demonstrating the existence of things out of the mind, that is to say, the reality of the extramental world. Whereas before your concessions, this possibility was ruled out.

PHILONOUS. I should agree with such a possibility. But you know how long a stretch there is between possibility and actuality.

HYLAS. Let us try. You have come to agree that reason is not passive and not confined to coordinating sense data, but is also able to create abstract ideas, and theories containing such abstract ideas.

PHILONOUS. I do.

HYLAS. Now, some of those theories, many of them, are designed to account for experience. Thus, it is a fact that there are theories of matter, of life, of mind, and even theories of theories.

PHILONOUS. Yes. But why could reality not be a product of theoretical activity?

HYLAS. No, you cannot envisage the possibility of converting to objective idealism. It is true that every true theory enriches the previously existing reality. But not every theory is true.

PHILONOUS. It is a truism that the number of wrong theories is by far greater than that of true theories.

HYLAS. And *that* is just one of my theoretical proofs of the existence of an independent outer world. First, if to think were the same as to exist, most people would not exist. Second, error among the few chosen ones would be unknown, and everybody would be a sage.

PHILONOUS. I must own that that is not the case.

HYLAS. This lack of complete overlapping or harmony between thoughts and things; this fact that disagreement between thinking and its objects is more frequent than the corresponding agreement, suffices to prove that thought is not the same as matter. That there is a reality, existing out of the mind, and which we are pleased to call 'matter'.

PHILONOUS. I never expected to see unsuccessful theories of matter used to prove the reality of matter.

HYLAS. In so far as our theories of matter fail, they thereby demonstrate the reality of matter; and in so far as they succeed, they demonstrate that we are able to understand the world surrounding us.

PHILONOUS. I must confess that the mere applicability of the concepts of truth and error proves that reality and its theoretical representations are not identical.

HYLAS. Then, it seems to me that I have succeeded in destroying your basic principles one by one.

PHILONOUS. I own that you have. I am now convinced that experience is not the sole source of knowledge, and that there are things beyond our perceptions, images and concepts. Only one question now troubles my soul: What has become of the omnipresent, eternal Mind, which knows and comprehends all things?

HYLAS. If my memory is still faithful after so many years, your favorite argument ran as follows: I know by experience that there are other minds exterior to my own; and, since every thing exists in some mind, there must be a Mind wherein all minds exist.

PHILONOUS. Exactly.

HYLAS. But you have agreed today that some things exist *out* of mind, so the very basis of your argument is gone. As for your argument based on the supposed passivity of the mind, which passivity would require an external mover, it fell long ago, as soon as you conceded that minds can create new objects of their own.

PHILONOUS. You have satisfied me, Hylas. I acknowledge it, and I think I shall withdraw until some next centenary.

SOURCES

Chapter 4 is based on my contribution to the Entretiens de Varna, of the Institut International de Philosophie, contained in Ch. Perelman, (ed.), *Dialectics/Dialectique* (The Hague: Martinus Nijhoff, 1975). Chapter 6 is a version of my paper in *Brain and Mind*, Ciba Foundation Series No. 69 (Amsterdam: Excerpta Medica, 1979). Chapter 10 appeared in a somewhat different form in the *Journal of Philosophical Logic* 3, 195–210 (1974). And the Appendix was first published in *Philosophy and Phenomenological Research* XV, 192–199 (1954). My warm thanks to all of the editors and publishers involved.

BIBLIOGRAPHY

Alexander, Samuel, (1920), *Space, Time, and Deity*, 2 vols. New York: Humanities Press.

Aristotle, *Categories*. In R. McKeon, (ed.), *The Basic Works of Aristotle*. Random House, 1941, New York.

Barash, D. P., (1976), 'Some evolutionary aspects of parental behavior in animals and man'. *Am. J. Psychology* **89**, 195–217.

Bindra, Dalbir, (1976), *A Theory of Intelligent Behavior*. New York: John Wiley & Sons.

Bindra, Dalbir, ed., (1980), *The Brain's Mind*. Gardner Press. New York.

Bohm, David, (1957), *Causality and Chance in Modern Physics*. Routledge & Kegan Paul, London.

Born, Max, (1949), *Natural Philosophy of Cause and Chance*. Clarendon Press, Oxford.

Bullock, Theodore, (1958), 'Evolution of neurophysiological mechanisms'. In A. Roe and G. G. Simpson, (eds.), *Behavior and Evolution*, pp. 165–177. Yale University Press, New Haven.

Bunge, Mario, (1955), 'Strife about complementarity'. *British Journal for the Philosophy of Science* **6**, 1–12, 141–154.

Bunge, Mario, (1959), *Causality*. Cambridge, Mass.: Harvard University Press. 3rd rev. ed.: Dover Publ. 1979, New York.

Bunge, Mario, (1967a), *Foundations of Physics*. Springer-Verlag, Berlin-Heidelberg-New York.

Bunge, Mario, (1967b), *Scientific Research*, 2 volumes. Springer-Verlag, Berlin-Heidelberg-New York.

Bunge, Mario, (1973a), *Method, Model and Matter*. D. Reidel Publ. Co., Dordrecht.

Bunge, Mario, (1973b), *Philosophy of Physics*. D. Reidel Publ. Co, Dordrecht.

Bunge, Mario, (1974a), *Sense and Reference*. D. Reidel Publ. Co., Dordrecht.

Bunge, Mario, (1974b), *Interpretation and Truth*. D. Reidel Publ. Co, Dordrecht.

Bunge, Mario, (1977a), *The Furniture of the World*. D. Reidel Publ. Co., Dordrecht-Boston.

Bunge, Mario, (1977b), 'Levels and reduction'. *American J. Physiology: Regularory, Integrative and Comparative Physiology* **2**, 75–82.

Bunge, Mario, (1977c), 'Emergence and the mind'. *Neuroscience* **2**, 501–510.

Bunge, Mario, (1979), *A World of Systems*. D. Reidel Publ. Co., Dordrecht-Boston.

Bunge, Mario, (1980), *The Mind-Body Problem*. Pergamon Press Ltd., Oxford-New-York.

Bunge, Mario, (1981), 'From mindless neuroscience and brainless psychology to neuropsychology'. *Trends in Neurosciences* **4**.

Campbell, Donald T., (1974), 'Evolutionary epistemology'. In P. A. Schilpp, (ed.), *The Philosophy of K. R. Popper*, I, pp. 413–463. Open Court, La Salle, Ill.

Cassirer, Ernst, (1965), *Determinism and Indeterminism in Modern Physics*. Transl. O. T. Benfey. Yale University Press, New Haven.

Cornman, James W., (1971), *Materialism and Sensations*. Yale University Press, New Haven and London.

Craik, Kenneth J. W., (1966), In S. L. Sherwood, (ed.), *The Nature of Psychology*. University Press, Cambridge.

Dimond, S. J., (1977), 'Evolution and lateralization of the brain: concluding remarks'. *Ann. New York Academy of Sciences* **299**, 477-501.

Dobzhansky, Theodosius, (1955), *Evolution, Genetics and Man*. John Wiley, New York.

Eccles, John C., (1951), 'Hypotheses relating to the mind-body problem'. *Nature* **168**, 53–64.

Eccles, John C., (1977), 'Evolution of the brain in relation to the development of the self-conscious mind'. *Ann. New York Academy of Sciences* **299**, 161–179.

Eccles, John C., (1978), 'Keynote address to the 1975 ICUS Congress', in *What ICUS is*. International Cultural Foundation (Rev. Myung Moon, Chairman). New York.

Eccles, John C., (1980), *The Human Psyche*. Springer International, New York.

Engels, Friedrich, (1878), *Anti-Dühring*. Lawrence & Wishart, 1955, London.

Engels, Friedrich, (1872–82), *Dialectics of Nature*, transl. C. Dutt. Lawrence & Wishart, 1940, London.

Frankfurt, H., Frankfurt, H. A., Wilson, J. A., and Jacobsen, T., (1946), *Before Philosophy: The Intellectual Adventure of Ancient Man*. Penguin, 1949, London.

Gonseth, Ferdinand, (1938), *La méthode axiomatique*. Gauthier-Villars, Paris.

Goslan, D. A. (ed.), (1969), *Handbook of Socialization Theory and Research*. Rand-McNally, Chicago.

Gruber, H. E. and Barrett, P. H., (1974), *Darwin on Man. Together with Darwin's Early and Unpublished Notebooks*. Dutton, New York.

Harlow, H. F., (1958), 'The evolution of learning'. In A. Roe and G. G. Simpson, (eds.), *Behavior and Evolution*, pp. 269–290. Yale University Press, New Haven.

Harris, Marvin, (1979), *Cultural Materialism*. Random House, New York.

Hartmann, Nicolai, (1957), 'Hegel und das Problem der Realdialektik', in *Kleinere Schriften*, Vol. II. W. de Gruyter, Berlin.

Hasenjaeger, Gisbert, (1966), 'Logik und Ontologie'. *Studium Generale* 19: 136–140.

Hebb, Donald O., (1949), *The Organization of Behavior*. John Wiley, New York.

Hegel, G. W. F., (1816), *Science of Logic*, transl. W. H. Johnston and L. G. Struthers, 2 vols. Allen & Unwin, 1929, London.

Hegel, G. W. F., (1830), *Encyklopädie der philosophischen Wissenschaften im Grundrisse*. Meiner, 1969, Hamburg.

Hodos, W. and Campbell, C. B. G., (1969), 'Scala naturae: why there is no theory in comparative psychology'. *Psychological Review* 76, 337–350.

Jaynes, Julian, (1976), 'The evolution of language in the late Pleistocene'. *Ann. New York Academy of Sciences* 280, 312-325.

Jerison, H. J., (1973), *Evolution of the Brain and Intelligence*. Academic Press, New York.

Kraft, Viktor, (1970), *Mathematik, Logik und Erfahrung*, 2nd ed. Springer-Verlag, Wien-New York.

Lange, Friedrich Albert, (1905), *Geschichte des Materialismus*, 2nd ed. Philipp Reclam jun., Leipzig.

Lashley, Karl S., (1949), 'Persistent problems in the evolution of mind'. *Quarterly Review of Biology* 24, 28–42.

Leakey, R., and Lewin, R., (1977), *Origins*. Dutton, New York.

Lenin, Vladimir Ilich, (1914–16), *Philosophical Notebooks*. Lawrence & Wishart, 1962, London.

Levy, J., (1977), 'The mammalian brain and the adaptive advantage of cerebral asymmetry'. *Ann. New York Academy of Sciences* 299, 264–272.

Llinás, Rodolfo, (ed.), (1969), *Neurobiology of Cerebellar Evolution and Development*. American Medical Association, Chicago.

Lloyd, G. E. R., (1966), *Polarity and Analogy*. University Press, Cambridge.

Malcolm, Norman, (1973), 'Thoughtless brutes'. *Proceedings and Addresses of the American Philosophical Association* 46, 5–20.

Masterton, R. B., Campbell, C. B. G., Bitterman, M. E., and Hotton, N., (eds.), (1976), *Evolution of Brain and Behavior in Vertebrates*. Erlbaum, Hillsdale, New York.

Masterton, R. B., Hodos, W, and Jerison H. J., (eds.), (1976), *Evolution, Brain and Behavior. Persistent Problems*. Erlbaum, Hillsdale, New York.

McMullin, Ernan, (ed.), (1964), *The Concept of Matter*. University of Notre Dame Press, Notre Dame.

Miró Quesada, Francisco, (1972), 'Dialéctica y recoplamiento'. *Dianoia* 1972, 182–198.

Moisil, Grigore, (1971), 'Les états transitoires dans les circuits séquentiels'. In
    O. Bîscă et al., Logique, Automatique, Informatique, pp. 215–268.
    Académie de la République Socialiste de Roumanie, Bucharest.
Munn, N. L., (1971), The Evolution of the Human Mind. Houghton & Mifflin,
    Boston.
Nagel, Ernest, (1961), The Structure of Science. Harcourt, Brace & World,
    New York.
Narski, I. S., (1973), Dialektischer Widerspruch und Erkenntnislogik. VEB
    Deutscher Verlag, Berlin.
Parker, Sue Taylor, and Kathleen Rita Gibson, (1979), 'A developmental
    model of the evolution of language and intelligence in early hominids'.
    Behavioral and Brain Sciences 2, 367–381.
Pawelzeig, Gerd, (1970), Dialektik der Entwicklung objektiver Systeme.
    Deutscher Verlag der Wissenschaften, Berlin.
Peirce, Charles S, (1955), In J. Buchler, (ed.), Philosophical Writings. Dover,
    New York.
Piaget, Jean, (1976), Le comportement, moteur de l'évolution. Gallimard,
    Paris.
Popper, Karl R., (1935), Logik der Forschung. Julius Springer, Wien. Rev.
    ed., The Logic of Scientific Discovery. Hutchinson, 1959, London.
Popper, Karl R., (1940), 'What is dialectic'? Mind N.S. 49, 403–426.
Popper, Karl R., (1945), The Open Society and its Enemies, 2 vols. George
    Routledge & Sons, London.
Popper, Karl R., (1953), 'A note on Berkeley as precursor of Mach'. British
    Journal for the Philosophy of Science 4, 26–36.
Popper, Karl. R., (1967) 'Quantum mechanics without "the observer"'. In M.
    Bunge, (ed.), Quantum Theory and Reality, pp. 7–44. Springer-Verlag,
    Berlin-Heidelberg-New York.
Popper, Karl R., (1968), 'Epistemology without a knowing subject'. In B.
    Van Rootselaar en J. F. Staal, (eds.), Logic, Methodology and Philosophy
    of Science III, pp. 333–373. North-Holland, Amsterdam.
Popper, Karl R., (1972), Objective Knowledge. Clarendon Press, Oxford.
Popper, Karl R., (1974), 'Intellectual Autobiography'. In P. A. Schilpp, (ed.),
    The Philosophy of Karl Popper, Book I, pp. 3–181. The Open Court Publ.
    Co., La Salle, Ill.
Popper, Karl R. and Eccles, John C., (1977), The Self and its Brain. Springer
    International, New York.
Putnam, Hilary, (1975) Mathematics, Matter and Method. University Press.
    Cambridge.
Quine, W. V., (1969), Ontological Relativity and Other Essays. Columbia
    University Press, New York.
Schneirla, T. C., (1949), 'Levels in the psychological capacities of animals'. In

R. W. Sellars *et al.*, (eds.), *Philosophy for the Future*, pp. 243–286. Macmillan, New York.

Scholz, Heinrich, (1941), *Metaphysik als strenge Wissenschaft*. Staufe-Verlag, Köln. Reprint: Wissenschaftliche Buchhandlung, 1965, Darmstadt.

Sellars, Roy Wood, (1922), *Evolutionary Naturalism*. Open Court Publ. Co., Chicago.

Smart, J. J. C., (1963), *Philosophy and Scientific Realism*. Routledge & Kegan Paul, London.

Stiehler, Gottfried, (1967), *Der dialektischer Widerspruch*. 2nd ed. Akademie-Verlag, Berlin.

Sussmann, Héctor, (1975), 'Catastrophe theory'. *Synthese* **31**, 229–270.

Tarski, Alfred, (1944), 'The semantic conception of truth and the foundations of semantics'. *Philosophy and Phenomenological Research* **4**, 341–375.

Thom, René, (1975), *Structural Stability and Morphogenesis*. W. A. Benjamin, San Francisco.

Thompson, R. F., (1975), *Introduction to Physiological Psychology*. Harper & Row, New York.

Vollmer, Gerhard, (1975), *Evolutionäre Erkenntnistheorie*. S. Hirzel, Stuttgart.

Woodger, J. H., (1929), *Biological Principles*. New ed., Routledge & Kegan Paul, 1967, London.

Zinov'ev, A. A., (1973), *Foundations of the Logical Theory of Scientific Knowledge (Complex Logic)*, rev. enlarged ed. Reidel Publishing Co., Dordrecht.

INDEX OF NAMES

INDEX OF SUBJECTS

OTHER BOOKS BY MARIO BUNGE

Temas de educación popular (Buenos Aires, El Ateneo, 1943).
La edad del universo (La Paz, Laboratorio de Física Cósmica, 1955).
Metascientific Queries (Springfield, Ill., Charles C. Thomas, 1959).
Causality (Cambridge, Mass., Harvard University Press, 1959; 3rd rev. ed.:
  New York, Dover, 1979). Spanish, Russian, Polish, Hungarian, Italian, and
  Japanese translations.
La ciencia (Buenos Aires, Siglo Veinte, 1960; 3rd ed. 1976).
Etica y ciencia (Buenos Aires, Siglo Veinte, 1960; 3rd ed. 1976).
Cinemática del electrón relativista (Tucumán, Universidad Nacional de
  Tucumán, 1960).
Intuition and Science (Englewood Cliffs, N.J., Prentice-Hall, 1962). Spanish
  and Russian translations.
The Myth of Simplicity (Englewood Clifss, N.J., Prentice-Hall, 1963).
Foundations of Physics (New York, Springer-Verlag, 1967).
Scientific Research, 2 volumes (New York, Springer-Verlag, 1967). Spanish
  transl.
Teoría y realidad (Barcelona, Ariel, 1972). Portuguese translation.
Method, Model and Matter (Dordrecht, Reidel, 1973).
Philosophy of Physics (Dordrecht, Riedel, 1973). French, Russian, and
  Spanish transl.
Sense and Reference (Dordrecht, Reidel, 1974). Portuguese translation.
Interpretation and Truth (Dordrecht, Riedel, 1974). Portuguese translation.
The Furniture of the World (Dordrecht, Reidel, 1977).
A World of Systems (Dordrecht, Reidel 1979).
The Mind-Body Problem (Oxford, Pergamon, 1980).
Epistemología (Barcelona, Ariel, 1980). Portuguese and German translations.
Ciencia y desarrollo (Buenos Aires, Siglo Veinte, 1980). Portugese translation.

EPISTEME

A SERIES IN THE FOUNDATIONAL, METHODOLOGICAL,
PHILOSOPHICAL, PSYCHOLOGICAL, SOCIOLOGICAL,
AND POLITICAL ASPECTS OF THE SCIENCES, PURE AND APPLIED

Editor: MARIO BUNGE
Foundations and Philosophy of Science Unit, McGill University

1. William E. Hartnett (ed.), *Foundations of Coding Theory*. 1974, xiii + 216 pp.
   ISBN 90–277–0536–4.
2. J. Michael Dunn and George Epstein (eds.), *Modern Uses of Multiple-Valued Logic*.
   1977, vi + 338 pp. ISBN 909–277–0474–2.
3. William E. Hartnett (ed.), *Systems: Approaches, Theories, Applications*.
   Including the Proceedings of the Eighth George Hudson Symposium, held at
   Plattsburgh, New York, April 11–12, 1975. 1977, xiv + 202 pp. ISBN 90–277–0822–3.
4. Wladyslaw Krajewski, *Correspondence Principle and Growth of Science*. 1977, xiv +
   138 pp. ISBN 90–277–0770–7.
5. José Leite Lopes and Michel Paty (eds.), *Quantum Mechanics, A Half Century Later*.
   Papers of a Colloquium on Fifty Years of Quantum Mechanics, held at the University
   Louis Pasteur, Strasbourg, May 2–4, 1974, x + 310 pp. ISBN 90–277–0784–7.
6. Henry Margenau, *Physics and Philosophy: Selected Essays*. 1978, xxxviii + 404 pp.
   ISBN 90–277–09001–7.
7. Roberto Torretti, *Philosophy of Geometry from Riemann to Poincaré*, 1978, xiv + 459 pp.
   ISBN–90–277–0920–3.
8. Michael Ruse, *Sociobiology: Sense or Nonsense?* 1979, xiv + 231 pp.
   ISBN 90–277–0943–2.
9. Mario Bunge, *Scientific Materialism*, 1981, xiv + 224 pp. ISBN 90–277–1304–9.